Twenty-First Century Biopolitics

BEYOND HUMANISM: TRANS- AND POSTHUMANISM
JENSEITS DES HUMANISMUS: TRANS- UND POST-HUMANISMUS

Edited by / Herausgegeben von Stefan Lorenz Sorgner

VOL./BD. 6

*Zu Qualitätssicherung und Peer Review
der vorliegenden Publikation*

Die Qualität der in dieser Reihe
erscheinenden Arbeiten wird vor der
Publikation durch den Herausgeber der
Reihe geprüft.

*Notes on the quality assurance and peer
review of this publication*

Prior to publication, the quality of the
work published in this series is reviewed by the editor of the series.

Bogdana Koljević

Twenty-First Century Biopolitics

PETER LANG
EDITION

Bibliographic Information published by the Deutsche Nationalbibliothek
The Deutsche Nationalbibliothek lists this publication in the Deutsche Nationalbibliografie; detailed bibliographic data is available in the internet at http://dnb.d-nb.de.

Cover image:
© Amorphogenesis by Jaime del Val -
Metabody Forum 2014 Madrid - Photo by Tais Bielsa

Library of Congress Cataloging-in-Publication Data
Koljevic, Bogdana, 1979-
  Twenty-first century biopolitics / Bogdana Koljevic.
    pages cm – (Beyond Humanism : Trans- and Posthumanism ; Vol./Bd. 6)
  ISBN 978-3-631-65977-9 – ISBN 978-3-653-05438-5 (e-Book) 1. Bio-politics–Philosophy. 2. World politics–21st century. I. Title.
  JA80.K655 2015
  320.01–dc23
                    2015009240

ISSN 2191-0391
ISBN 978-3-631-65977-9 (Print)
E-ISBN 978-3-653-05438-5 (E-Book)
DOI 10.3726/ 978-3-653-05438-5

© Peter Lang GmbH
Internationaler Verlag der Wissenschaften
Frankfurt am Main 2015
All rights reserved.
Peter Lang Edition is an Imprint of Peter Lang GmbH.

Peter Lang – Frankfurt am Main · Bern · Bruxelles · New York ·
Oxford · Warszawa · Wien

This publication has been peer reviewed.

www.peterlang.com

For Nikola, Milica and Srđan

Special Acknowledgement

The author wishes to express profound gratitude to the Studenica Foundation, for recognizing the philosophical and cultural importance of this work and contributing financially to make its publication possible.

Content

I

The Phenomena Of Contemporary Biopolitics *vs*. Potentialities For Rethinking The Political

What Is Contemporary Critique
Of Biopolitics?

To begin with, a political-philosophical analysis of biopolitics in the twenty-first century as its departure point, suggests the difference between Foucault's theory and contemporary investigations, particularly Agamben's, Hardt's and Negri's reflections. This difference, mostly between such works as *The Birth of Biopolitics* and *Society Must Be Defended,* on the one hand, and *The History of Sexuality vs. Homo Sacer, State of Exception, Means Without End, Empire* and *Multitude,* on the other, leads to the *rethinking of the political* and, therefore, to the issue of *political subjectivities.* Moreover, Foucault's *genealogy of the present* – as the path for conceptualizing the multiplicity of appearances of contemporary biopolitics – actually leads to the rethinking not only of politics, but also of ethics and law. In comparison with *post-political* theories, the project of genealogical critique presents perspectives and potentialities for the political practice *beyond biopolitics.* Doubtlessly, Foucault does not in any case endorse either moral or political ends, and yet his *oeuvre* outlines what a *political philosophy,* faithful to his thinking, would look like – and that it could not be reduced to *microphysics of power.*

Our thesis here, therefore, is twofold: the basis for comprehending contemporary phenomena of biopolitics lies precisely in Foucault's theory of *biopolitics as neoliberalism* whereas the response to such a situation can also be found perhaps, in a rethinking of freedom as a new potentiality of democratic politics. This is to say that an analysis of the birth of biopolitics in history of liberalism aims not only at presenting a specific type of governmentality – which significantly differs from all the rest by its instrumentalization of life as such – but also at manifesting *the potentialities of genealogy for rethinking the political.* In short, Foucault's philosophical investigations of the history of biopolitics present the issue of the political and suggest exemplary prolegomena for a political philosophy and political practice. Likewise, this presupposes the insight that the conceptualization of power and/or biopower, together with the whole discourse on microstrategy of power, is not Foucault's final word, i.e. that the disappearance of the political, seemingly paradoxically, is not total.

What emerges here is rather some room for political subjectivity as a possible area of freedom.

This is one of the most original and extraordinary consequences of Foucault's theory which, *en generale*, has not sufficiently been taken into account in contemporary philosophy. This implies a perception of political genealogy as, in all its relevant aspects, a project of creative, empirical and local critique of multiplicities of contemporary phenomena of biopolitics. It also implies the need to articulate the genealogy of the present as a meeting point of the struggle between power and subjectivity so that we should rethink an alternative conception of politics and power. This is why, if philosophy attempts to articulate the phenomena of biopolitics in the twenty-first century, an inquiry into this relation appears a necessary task.

Indisputably, radical politics cannot endorse any sort of "minimalism" of power relations, i.e. any claim which implies that power games can be played with "minimum of domination" (Foucault), but this does not mean that we should dismiss other significant implications of Foucault's theory, especially for democratic politics. Firstly, in his analysis of biopolitics as the discourse and politics of neoliberalism (*The Birth of Biopolitics*), Foucault emphasizes *the relation between theory and politics*, and reflects on *conditions of possibility of political philosophy*. In this sense, if a theory attempts to articulate itself as political philosophy, its beginning lies in a critical examinations of *liberal power politics*. Or, as it has been emphasized ever since *The History of Sexuality*, biopower has been the *sine qua non* for the development of capitalism, and this is precisely the point from which the analysis begins. This is the reason why biopolitics – through its phenomena of the market, liberal economy and multiple techniques of governing – appears as the *practice of the truth* of liberalism, i.e. as the power of anti-politics which leads to the re-thinking of the political. Secondly, the question is *how does politics happen according to Foucault* and, particularly, why is *the state* not "a cold-blooded monster" and, therefore, not the ultimate power of biopolitics. Finally, why is it that Agamben, on the one hand, and Hardt and Negri on the other – both in difference to Foucault – recognize the state as the *par excellence* enemy, i.e. as the irreplaceable carrier of biopolitics?

In this context, it is worthwhile to perceive the structural similarities – both theoretical and political – between Agamben's theory and Hardt's and

14

Negri's theories. The first proximity lies in arguing for political subjectivity as a *"singularity-commonality"* – both in Agamben's conception of the so-called "whatever singularities" and community of such singularities, and in the Hardt's and Negri's ideas of the relation between "the singular" and "the common". Both theories are highly opposed to the state in all its aspects, while the silent impulses and the outlines of these contemporary discourses can be traced to the indisputable influence of globalization and postmodernism. Our primary concern here is that such *post-state theories* simultaneously dismiss concepts such as *"sovereignty"* and *"the people"*; moreover, we see their decisive persuasion that these concepts must be superseded *at any cost.* In Agamben, it is precisely the identification of *biopolitics* with *sovereignty* which is the leading motive for the appeal for a "non-community" of "whatever singularities", articulated so as to replace the concept of "the people"; while in Hardt and Negri it is at first ambivalence, and then their rejection of sovereignty and the people in the multiplicities of singularities of the so-called multitude.

In this sense, in *The Coming Community, The State of Exception, Means Without End, Empire* and *Multitude* the ideas of resistance to "the Empire" or to "the state of exception" are articulated in a way which suggests that "the coming community" or "the multitude" necessarily stand against the power of biopolitics, i.e. against the power of the state. Or, individuality, singularity and multiplicity in these discourses appear as ultimate carriers of a new political subjectivity opposed to sovereignty and the people. Consequently, the second similarity between two contemporary theories of biopolitics is their relation towards law, i.e. the emphasis on insufficiencies and inefficiencies of both national and international legal systems. The leading trace in these critiques is the role of sovereignty as that which must be superseded so that "the new humanity" can properly appear. Moreover, this typically postmodern gesture is accompanied by key arguments which rest on the globalization processes they attempt to oppose.

In short, in Hardt and Negri's theories it is precisely the central thesis about the multitude which is encouraged, and even produced by the system, while in Agamben the emphasis placed on such issues as the territoriality of a non-state of refugees, which emerges from politics and discourses of global individualism, i.e. from contemporary liberal governmentality as

15

biopolitics *par excellence*. From such presuppositions, Agamben's *post-ethics* of new humanity – and at the same time *post-politics* – is developed as a theory of singularity, individuality and non-responsibility.

The theories about the multitude and about refugees, therefore, in their analysis of the political and biopolitics, lack precisely the response to the issue of *contemporary phenomena of biopolitics*. They do not respond to the question *how does the politization of life happen at the present time*. How is contemporary neoliberal governmentality exercising power over life? What are the political forms though which the disappearance of political subjectivity is realized? What is happening with democracy in leading discourses and dominant political practices? Does the political need to be removed in order to open the space for the humanitarian? What is the relation between sovereignty and liberalism in the West today? Is not Foucault's argument about biopolitics as a strategic relation between knowledge and power, which takes place in the process of fragmentation and dissolution of sovereignty, even more persuasive in the twenty-first century? Or, more precisely, if we rethink contemporary neoliberalism in its attempts to annihilate political sovereignty and legal sovereignty, is it not the case that Foucault's elaborations sound practically prophetic today?

Certainly, we can say Agamben's theory, as well as theories of Hardt and Negri, lack material, i.e. empirical political insights into contemporary power politics and the entire sphere of *Realpolitik*, insights into different military interventions in the world, insights into the liberal doctrine of interventionism, so that the relation between the political and the humanitarian for the most part remains beyond these conceptions. Secondly, it seems there is also a philosophical lack present in such theories. The question is how has biopolitics depoliticized politics and what does this have to with popular sovereignty, democracy and ethics? To throw these questions on the table also means to perceive the third similarity between Agamben's theory and Hardt and Negri's theories, namely, their *sameness* in relation to Foucault's theory of biopolitics followed by a more or less silent abandoning of such a project. This becomes clear if one attempts to address political subjectivity and the contemporary phenomena of biopolitics. The structural difference here relates to the influences of postmodernism and liberalism in the biopolitical theories of Agamben and

Hardt and Negri, i.e. in Foucault's claim that biopower is, to a very great extent, opposed to this.

It is most extraordinary that Agamben's reflections, ultimately, present a bond between metaphysics and politics (*Homo Sacer – Sovereign Power and Bare Life, Means Without End*), while in Hardt's and Negri's discourse this is expressed as the relation between ontology and politics. This is explicit in the idea of the so-called "first multitude", i.e. the "ontological multitude", as a constitutive principle of the second, derived, "political multitude" (*Multitude*). On the other hand, in Agamben this bond becomes clear in the construction of metaphysics of sovereignty or, more accurately, in the transcendence which enables a different transcendence of "potentiality of potentialities" belonging to a post-ethical subjectivity. The transcendence of the sovereign, the description of sovereignty as nothing but a *"pure form* of power"[1], makes possible Agamben's metaphysics which ends in the quest for *"pure humanity"*. Moreover, by such a movement Agamben accepts the contemporary liberal idea about "the end of history", presenting this metaphysical potentiality as the exclusive (post)-messianic way out, in the uncanny resemblance between what he finds intolerable and his view of the desired destination for the so-called "happy life". Paradoxically, therefore, "the sovereign ban" leaves the subject in the state of "bare life" which is suspiciously similar to the desired "whatever" condition of human existence. Such discourse manifests itself as (post)-messianic suspension of complete history of Western politics and philosophy, and is exemplified in the idea of "the end of history" as "the end of state".[2] Or, in Agamben's own words, what we are dealing with here is the much redeemed post-messianic existence in which "everything will be as is now, *just a little different.*"[3]

On the other hand, subjectivity in Foucault has nothing to do with substantialization, i.e. subjectivity rises at the heart of intersubjectivity: it is imminent, relational and empirical. The difference between "subjectivity"

---

1   G. Agamben, *Homo Sacer – Sovereign Power And Bare Life,* Stanford University Press, Stanford, 1998, p. 26.
2   *Ibid.*, p. 60.
3   G. Agamben, *The Coming Community,* The University of Minnesota Press, Minnesota, 1993, p. 57.

and "singularity" consists in the fact that the latter is practically inter-changeable with "individuality", which explains the use of both concepts (singularity/individuality) in liberal discourses. In this sense, in globaliza-tion discourses the so-called *cosmopolitan individuality* has occupied a special place, being instrumentalized to the point of clashing with *political subjectivity*, inasmuch as this is the term more and more used against *bi-opolitics*, i.e. *neoliberalism*. The crossroad here rests in Foucault's percep-tion of the specificity of neoliberalism as biopolitics: its governmentality does not consist in the power over the citizen as a legal subject but in the power over the citizen inasmuch as he appears as a part of biomass called *population*; therefore, it is not concerned with legality, as Agamben, Hardt and Negri argue, but precisely with *life*. Foucault continues in expressing how governmentality needs to be analyzed outside the model of Leviathan, i.e. outside the field of legal sovereignty, because it rests on *techniques of domination* (*Society Must Be Defended*). Making it clear that the new forms of governmentality *colonize* legal structures and *dissolve the legal system of sovereignty* (*Society Must Be Defended*), Foucault tries to sug-gest that biopolitics refers to politization of life of individuals and popula-tions, and in this way has more to do with techniques of domination that develop *beyond the sphere of institutions and law*.

In this context, according to Foucault, the issue of the *measure of gov-erning* as well as the issue of subjectivity, namely, as the question "Who are you?" need to be addressed. What Foucault calls, for example, "*the move-ment from the body to population*", also deals with the same phenomenon of contemporary biopolitics, which refers to the processes of *natality* and *mortality*, as well as to "*the problem of the city*", outlining all appearances of the movement from *control* to *regulation*.[4] Moreover, in spite of the fact that Foucault does not endorse political sovereignty as *popular sov-ereignty*, i.e. as a discourse which presents *a response* to biopolitics – this

---

4   Foucault emphasizes that most contemporary forms of power in themselves contain both moments, exemplifying this with sexuality: sexuality simultane-ously refers to political anatomy of human body and to biopolitics of popu-lation. The second example is Nazism, in which both power of control and biological regulation were equally present.

is a political and ethical demand of the type Foucault refuses to make – he nonetheless sketches the possibility of such a project.

Foucault's differentiation between *"revolutionary course and law"* and *"utilitarianism and liberal state practice"* (*The Birth of Biopolitics*) represents, ultimately, the difference between *rethinking of popular sovereignty vs. the power of biopower.* Moreover, this is the perspective from which Foucault approaches the issue he perceives as the political, i.e. the issue of freedom, since the specificity of *positive freedom*, in opposition to *negative freedom*, is reflected in the battle between "the revolutionary course" and "the utilitarian course". Ultimately, the first presents a potentiality of the *power of freedom,* while the second relates to *freedom of power.* The phenomena of the market, sexuality, prison, madness etc., as depolitization proper – as the power "to live and let die" – belong to "the utilitarian course" of liberal tradition as biopolitics. The structural difference between two courses reveals a potentiality of freedom and, as such, presents a possible basis for a political philosophy faithful to the task of genealogy of the present. Furthermore, while the "revolutionary course" moves from the discourse on human rights to the discourse on sovereignty, the "utilitarian course", in opposition, *is not based on law* but, rather, on state practices, and has "usefulness" as its final criterion instead of legitimization. According to the "revolutionary course", *law* arises from *collective will*, i.e. from the very idea of *the social contract,* whereas in utilitarianism *law* appears as a *result of transactions* which divide state power and the individual (the distinction which corresponds to the difference between "positive" and "negative" *freedom.*)

It is from the prevalence of such utilitarian, liberal thinking, that the techniques of governing developed, together with the biopolitical fracture, since it further enabled categories such as *population* to become *more relevant then legal concepts.*[5] According to Foucault, therefore, there has never been such a thing as substantial legal theory in liberalism, since liberalism undertook something completely different – the development of power throughout governing, *where legal subjectivity is arbitrary, a*

---

5   New "political rationality" of biopolitics is, therefore, significantly related to the development of empirical sciences, as a way of dismissing the language and the arguments of political philosophy and theory.

relative moment, and a moment which can in certain cases be *used*, and therefore instrumentalized. This is because the key player, and carrier, of liberalism, is the figure of *homo economicus*, and he cannot be reduced to a legal subject. Such movement clearly leaves *sovereignty and law* on one side, and *economy and liberalism* on the other. Moreover, Foucault writes that *"neither democracy nor the legal state were not necessarily liberal, nor was liberalism necessarily democratic, or faithful to legal norms."*[6] Thus the analysis of the contemporary phenomena of biopolitics opens new possibilities for democratic practice as well.

---

6   M. Foucault, *The Birth Of Biopolitics*, Palgrave Macmillan, New York, 2008, pp. 436–437.

From International Terrorism To International Institutions: Liberal Interventionism Or Post-Liberal Internationalism?

What are, after all, the phenomena of biopolitics in the present? What does a contemporary genealogy look like, since it seems that this genealogy appears as an exemplary question of political philosophy? The issue is how the *politization of life* is happening both in political discourses and politics *per se*. Consequently, such analysis will equally relate to potentialities of *political subjectivity*. Liberal power over life, in last decades of the twentieth century and at the beginning of the twenty-first century shows itself in a series of wars and interventions, but, no less significantly, in the preservation of *empty space* in respect to articulation of its actions. This way, its manifestation concerns not only the level of *Realpolitik*, but, perhaps even more, the level of conceptualization that has enabled and produced *post-politics*. Moreover, this is precisely what Foucault has called the *regime of truth* of biopolitics, as the instance of *thinking* in *power practice of dominance* – which he articulated as multiplicities of the manifestation of the *discourse of war*.

The bottom line of contemporary phenomena of biopolitics, therefore, lies in Foucault's observation that contemporary wars are now being led "*in the name of life*" – biopolitics proper – and "*in the name of peace*". Such wars are led for the biological survival of one population, in difference and opposition to the survival of the other population, which appears as *the enemy*. These wars are, most notably, manifested in the so-called *interventions* in numerous sovereign states, and in that sense differ from *international terrorism*, which appears simultaneously, but is a different phenomenon. Military interventions and terrorism, therefore, appear as phenomena of contemporary biopolitics *par excellence*, and moreover, precisely as such present traces for comprehending other phenomena which are no less relevant but are not so transparent and visible. This, first, refers to the relation between political, legal and military international institutions and alliances – and the doctrine of liberal interventionism as *humanitarianism*. In recent history, this bonding – as in the cases of Iraq,

Afghanistan and Serbia – has produced death of thousands of civilians in different countries, i.e. it has sacrificed *multiplicities of populations* in favor of the chosen ones, and, while officially appealing for *democracy* and *freedom* has mostly abused precisely these concepts. *Politicizing life* and *depoliticizing politics*, i.e. replacing it with *the humanitarian rhetoric,* this practice has significantly been one against sovereignty and, moreover, precisely against sovereignty in its aspects which stands for *autonomy* and *subjectivity.* This has become apparent in different spheres of politics, law, economy, science, philosophy, culture and everyday life.

But let us first turn to terrorism. It seems that most of interpretations – whether conservative, liberal or communitarian – have not provided a persuasive response to an obviously new situation in the contemporary world, the situation which began to emerge at the time of the attacks on Twin Towers in New York in 2001. On the one hand, what is at stake with terrorism is *self-sacrifice and sacrifice of life itself*, which happens in the very same act which attempts to manifest itself *as a political act.* But this has also to do with a partially paradoxical and deeply perplexed relationship between *terrorism* and the *system* it opposes. Baudrillard's claim that "terrorism, like a virus, is everywhere around us"[7] – as well as his controversial views expressed in *The Spirit of Terrorism* – discloses precisely the moment of relation and the moment of rupture between terrorism and the system as *its other*. For Baudrillard, terrorism presents a *return of the real* in the discourses of *"Western illusions"*, as the exemplary "anti-body", which simultaneously appears as *produced by the system* and as the final moment of its *rupture*. In this sense, Baudrillard's discourse on terrorism – in difference to simplified interpretations which dismiss it as "evil as such" – presents a relevant analysis which deeply considers this phenomenon, and refers to some implications of the neoliberal governmentality. Terrorism, as the *other of the system*, is its collateral effect. Simultaneously, there exists a relevant *excess* in terrorism as well, which consists in its movement *beyond* the system, i.e. at the point where its disruptive character emerges in full force. Furthermore, the emphasis here lies on the possibility of terrorism to be comprehended as a response to the depolitization

---

7 See J. Baudrillard, *The Spirit Of Terrorism*, Verso, New York, 2003.

of biopolitics – and while this does not justify it, it shows the complexity of such a phenomenon, for it suggests its philosophical location, between *politization* and *depolitization*, and at the same time reveals its causes in neoliberalism.

In other words, if one follows Foucault's idea that biopolitics, as neoliberalism, operates on the basis of governing over populations and realizes its power, as biopower, through different types of "*maximalization of life*"[8] which significantly contributes to growth of depolitization – then it is precisely *the minimalization of living*, indisputably manifested in terrorism, that *disturbs the existing order*. Or, in spite the fact that its act is one of violence, in the aspect of its *radical nihilism* in respect to *ideology of life,* terrorism, implicitly or explicitly, attempts to remind us of politics. Traces of such thinking are equally present in Badiou's *Polemics*, i.e. in his articulation of two, practically binary opposed paradigms: the first symbolizes the liberal system ("the materialistic paradigm") while the second presents *the nihilistic approach*, such as terrorism.[9]

In this light terrorism appears simultaneously as the act of depolitization and the act of politization, since it politicizes life for particular political purposes, i.e. it arises as the battle of one population against the other, expressed as the enemy – and is, therefore, the act of biopolitics *par excellence*. Moreover, its anti-political stance is equally presented in the *sacrifice of life of innocent others*, but in this respect it is structurally similar to the system it opposes. The difference, however, between liberal governmentality and terrorism, as its collaterally created excess – the aspect that refers to *traces of the political* in the case of the latter – lies in the described *ideology of maximalization of life*. In short, *self-sacrifice* is precisely the infinite otherness from the perspective of neoliberalism which finds it incomprehensible, since it rests *beyond the system*. Furthermore, *self-sacrifice* of one individual which belongs to one population, and then

---

8   "Maximalization of life" refers to a process which manifests itself in different aspects of *biopolitics of everyday life*: the social imperative of health care, the demand to be young, beautiful and to live longer, the ideology of anti-aging, the *positive discrimination* of woman, minorities, and persons with special needs, the promotion of GMO industry, present some of its most visible examples.

9   See A. Badiou, *Polemics*, Verso, New York, 2006.

of the second individual belonging to the same population etc., shows itself as the desperate impulse of *one entire population,* as the *dramatic act of resistance* to biopolitical governmentality *per se.*

This is why acts which consist exclusively in *self-sacrifice,* i.e. not in the sacrifice of others, in totalitarian times, manifest themselves not simply as acts of self-destruction, but also as acts which obtain a new quality – *as political acts proper.* Such is the case of self-immolation on Tahrir Square, and, more recently, the four cases of self-immolation in Bulgaria, which is a member of EU, but the country in which poverty; misery and hopelessness have reached the point of *bare life* as the only offered to the entire population. In other words, these dramatic and radical acts appear as acts of *refusal* of imposed political and economic situations; acts in the name of *human dignity,* which does not accept the order of *bare existence* as the only *value* and which distinguishes between *bare life* and *dignified life.*

But what have been the responses, first to terrorism (sacrifice of others and self-sacrifice), and then to these specific examples, such as self-immolation? The Bush administration launched the so-called *war* against *terrorism,* which, as it has been later articulated in the National Security Strategy of the USA (2006), is to be led in the name of "*advancing freedom's cause*". This way, US National Security Strategy explicitly authorized preventive use of military force against all "*enemies of freedom*", moreover, claiming that such war is necessary for *democracy.* Simultaneously, along the same lines, the US National Defense Authorization Act (NDAA), adopted in December 2011[10], permits American army to capture, imprison and hold for an indefinite period of time, and without the right for defense, *all* persons which are suspects (US citizens *included*), and this exemplifies an even further step of "development" of contemporary biopolitics.

---

10 Moreover, this law appears along the lines of continuity of contemporary US law in the last decade, as a specific, and certainly even more radical sequence of the US Patriot Act, voted out in September 2001, and of National Security Strategy of USA, from 2003, which to a major extent, in last instance, presents itself as *Realpolitik* of war.

Here Bernstein's analysis, presented in *The Abuse of Evil: The Corruption of Politics and Religion since 9/11,* appears as exemplary. Emphasizing the difference between *absolutistic thinking* and *democratic thinking,* Bernstein argues that total self-certainty is *per definitionem* a presupposition of war, while democratic politics in itself refers to the openness of questions about theory and knowledge. Analyzing the response to 9/11, he notes that the speech about "*absolute evil*" de-legitimized the discourse of "war against terrorism", since such articulation is anti-political in itself.[11] From this perspective, it is clear there is a structural contradiction in liberal theories, especially in Ignatieff's theory of the difference between the so-called "greater evil" and "the lesser evil". In *The Lesser Evil* Ignatieff calls terrorism the "greater evil" while *the lesser evil* is the liberal response to it, which is supposed to be justified *as a defense of the political.*[12] The *contradictio in adjecto* of such a theory lies both in the concept of what is to be taken as a "*liberal-democratic*" response to terrorism and in its concrete political practices. It is hardly a matter of dispute that terrorism to a great extent is anti-political – and that its becoming a rule i.e. its universalisation would lead to a pre-political state as a state of war (*bellum omnium contra omnes*). However, it is anything but plausible that the "liberal-democratic" response "by itself" has much to do with non-violence, dialogue and reason. Moreover, how can an approach which, *also,* and *significantly,* causes death of civilians and innocent people, be called *a lesser evil?*

If we turn to radical and dramatic examples, we can perceive more precisely how this structure works. If North Korea were to use thermo-nuclear force against any state this, doubtlessly, would be the case of the worst

---

11 See R. Bernstein, *The Abuse Of Evil: The Corruption Of Politics And Religion Since 9/11,* Polity Press, Malden, MA, 2005.

12 "Terrorism is an offense against politics itself, against the practice of deliberation, compromise, and the search for nonviolent and reasonable solutions. Terrorism is a form of politics that aims at the death of politics itself. For this reason, it must be combated by all societies that which to remain political: otherwise both we and the people terrorists purport to represent are condemned to live, not in a political world of deliberation, but in a pre-political state of combat, a state of war." M. Ignatieff, *The Lesser Evil,* Princeton University Press, 2004, pp. 110–111.

possible state-terrorism and a return to a pre-political state of war *per se.* But what can we say of the innocent killed in "wars *against* terrorism" or – which is also a specificity of a "liberal-democratic" approach – about the death of thousands of civilians as *"collateral damage"* of the so-called *"humanitarian wars"*? Where does this leave *politics, ethics* and *law*? Is the argument about *the lesser evil*, therefore, *a political one*? Moreover, Ignatieff's final response to the question why not stay on the side of international law is the following: "We are faced with *evil people.*"[13] Such crypto-moral interventionism, therefore, appears as the introductory force for military interventionism. This way, the liberal discourse established precisely the binary oppositions which Bernstein perceived as deriving from a quasi-Manichean order. i.e. oppositions which exemplify a structural similarity as *anti-politics.* Finally, it has unwillingly confirmed that the idea of *bellum justum* presents itself as a *contradictio in adjecto*, since the concept of democracy, as well as the concept of justice, cannot be expressed as co-belonging to the concept of war.

The key characteristic of contemporary biopolitical governmenality, which calls itself *liberal-democracy*, lies in what in Foucault's terminology would be called a *dispositive of security*, i.e. in the emphasis of *permanent threat* to one population, which is supposed to justify even the indefinite suspension of international law as the basis of seemingly peaceful international relations. This means that the dispositive of security rests in the heart of governmentality of regulation of population in multiple aspects. A different, equally relevant example of this regulation based on the security/threat mechanism has been articulated by Sarasin as *bioterror.* Sarasin shows that bioterror manifests itself as a phantasm in his analysis of *anthrax,* as the proper metaphor for the *anonymous threat.*[14] Sarasin argues that the terror that comes from inside is called anthrax, emphasizing, therefore, how unconscious messages spread through the imaginary ones, presenting bioterrorism as *the dark side* of the so-called globalization.

But let us turn once more to the ideology of *humanitarianism*, i.e. to its biopolitical relation to law and international institutions. In the golden age of triumph of neoliberal promotion of the so-called *human rights* as the

---

13 M. Ignatieff, *The Lesser Evil*, p. 12.
14 P. Sarasin, *Anthrax. Bioterror als Phantasma.* Suhrkamp, 2004.

ultimate measure of politics, the idea was not only to marginalize or completely dissolve existing key international institutions, but equally to establish new ones, exclusively reserved for the West. Such was the example of the appeal for constitution of the League of Democracies, in which a relatively small number of selected Western states would participate. As Gelb has emphasized, in advocating the idea of "a concert of democracies", it is relevant that "some of the liberal Democrats have joined with the neoconservatives", and that the agreement also presupposed that "they make little room in their concert for China and Russia."[15] Moreover, ideas such as one of League of Democracies or a concert of democracies, have rested precisely on the argument that "in dealing with civil wars, the legal and institutional framework of the United Nations seemed to provide not an aid, but an obstacle" for "such policies were not a fit target for international sanction, let alone military intervention."[16]

The moment of particular relevance here is on the special emphasis that "humanitarian intervention serves as a reminder that such measures must remain occasional and exceptional – for it remains an intervention: an infringement of state sovereignty whose legal, political and moral status remains to be clarified."[17] In such sense – and precisely in accordance with Foucault's articulation of biopolitics as dissolution and fragmentation of political sovereignty – contemporary biopolitical phenomena appear in structural relation to self-proclaimed exclusivity of the West and its attempts to institutionalize its own laws and declare them as ones belonging to the wholeness of the so-called international community. Last but not least, there exists also a conceptual link between terrorism and these discourses on the *inefficiency* of international institutions such as the UN, and simultaneous appeals for new Western institutions: it is precisely *the ideology of human rights*. Or, the humanitarian rhetoric enabled the possibility for certain terrorist actions to appear as justified and *vice versa*, such rhetoric served well in the explanation of the need for interventions,

---

15 L.H. Gelb, "Necessity, Choice And Common Sense", in *Foreign Affairs*, vol. 88, No. 3, May/June, 2009.

16 G. Reichberg, H. Syse, E Begby, *The Ethics Of War*, Blackwell Publishing, Oxford, 2006 p. 683.

17 *Ibid.*, p. 684.

as well as for the establishment of new exclusive institutions. This has been articulated by Tony Blair, as one of the leaders of the doctrine of interventionism in the following way: "Globalization is not just economic. It is also a political and security phenomenon...We need new rules for international co-operation and new ways of organizing our international institutions."[18] Moreover, this statement comes forth precisely from Blair's previous insistence that NATO 1999 Kosovo war was "a just war" for one "cannot let the evil stand" – an evil which can properly be described as "a threat to international peace and security."[19]

Here we can recall Foucault's observation that new biopolitical governmentality "colonizes legal procedures and destructs the legal system of sovereignty".[20] The radical example of the same issue is the situation of war becoming the code for peace, and this, according to works such as *Society Must be Defended* and *The Birth of Biopolitics,* happens through a discourse which remains necessarily alien to philosophers and lawyers. Moreover, as Foucault argues, what is at stake here is no longer a historical struggle, but a struggle that transforms itself from a historical struggle into a biological form of "struggle for life" i.e. for survival – where the leading idea is again one of a "better suited race" and selection of some in opposition to "others". Contemporary biopolitics, in this sense, as its leading trace, has precisely the articulation that "wars are now being led in the name of life": "the principle that it is possible to kill in order to live" thus becomes "the principle of international strategy", and survival now is "not the legal survival of sovereignty but the biological survival of a certain population."[21]

Moreover, precisely in this sense one more relevant phenomenon of contemporary biopolitics simultaneously emerges – the issue of political trials. In *A History of Political Trials* Laughland articulates that political trials – such as extradition of the former Peruvian president Alberto Fujimori

---

18 T.G. Ash, "Like it or loathe it, after 10 years Blair knows exactly what he stands for" *The Guardian*, 26 April 2007.
19 T. Blair, "Doctrine Of The International Community", Speech To The Economic Club of Chicago, Hilton Hotel, Chicago, April 22, 1999.
20 M. Foucault *Society Must Be Defended*, Picador, New York, 2003. p. 55.
21 *Ibid.*, p. 44.

in 2007, the trial of Jean Kambanda, the former prime minister of Rwanda in 1998 or the trial of Slobodan Milošević, former president of Yugoslavia from 2001–2006 – "are indeed part of a new trend towards military and juridical interventionism and towards rule by supranational political and juridical institutions."[22] Thus, Laughland argues, states become subjects to a body of law which is tailor-made for the purpose of controlling them. Or, as one of most notable proponents of such a position, Robertson has explicitly confirmed that "the movement for global justice has been a struggle against sovereignty."[23] Therefore, while the doctrine of human rights is constructed to appear as incontestably moral, and as such beyond politics, it is precisely the basis of a highly ambitious political project, which significantly involves the creation of a new supranational jurisdiction and new law, i.e. an establishment of a new and different *right to rule*, and in this sense, the right to *legitimize domination*. Such development has "inevitably led to the wielding of new political power through war."[24]

Moreover, new political and juridical institutions prosecute political actors as *criminals* and, therefore, appear as biopolitical *par excellence*. By reducing state acts to *private crimes* neoliberal governmentality has simultaneously concealed the aspect of legality of the rule and produced a new set of rules to replace existing principles. In this sense, Laughland claims that "the postwar international system, created out of the ruins of World War II, was based on the Nuremberg principles that starting a war is the supreme international crime, and on the concomitant principles of the sovereign equality of states and the rule against intervention in the internal affairs of other countries. NATO's attack on Yugoslavia was precisely intended to overthrow these rules and replace them with new principles which would permit what had previously been solemnly forbidden".[25]

---

22 J. Laughland, *A History Of Political Trials*, Peter Lang, Oxford, 2008, p. 13.
23 G. Robertson, *Crimes Against Humanity: The Struggle For Global Justice*, London, Penguin Press, 2000, p. 18.
24 J. Laughland, pp. 14–15. See also D. Chandler, *Bosnia: Faking Democracy After Dayton*, Pluto Press, London, 1999.
25 J. Laughland, p. 221. Moreover, Laughland emphasizes that "the alliance between NATO, the KLA, and the ICTY was therefore not just tactical or even strategic, it was ideological at the deepest level. NATO leaders based their arguments for war on their claim that, in the face of terrible atrocities, the rules

Finally, in *A History of Political Trials* the emphasis is placed on the issue that the history of political trials shows "zero per cent acquittal rate", which, consequently, brings us to Laughland's conclusion that political trials are a continuation of war by other means, i.e. that trials themselves appear not as juridical acts but as political acts – or, rather, *politicized acts* of *depolitization*.

These issues of political trials, juridical interventionism, military interventionism and political interventionism – based on the so-called *humanitarianism* – present the multiplicity of appearances of the friend/foe principle in theory and practice. Contemporary biopolitics proper, therefore, manifests itself as a liberal politization of life in different aspects. The juridical issue itself appears through numerous examples, while, indisputably, one of the most relevant cases in this respect is the twofold decision of the International Court of Justice on the legality of secession.[26] This was for the first time in history that such a matter was a subject of debate, and the *undecidablity* expressed in the opinion of the court[27] was precisely the result of the awareness of its extraordinary consequences for international law *en generale*.

In this sense, seemingly paradoxically, the relation between political subjectivity, political sovereignty, law and democracy arises as a potential

---

of national sovereignty had to give way to the right of intervention and the superior claims of human rights. This enabled them to claim that their attack was justified, since it was patently illegal under international law, having not been authorized by UN Security Council. NATO's anti-sovereign philosophy was identical to that of the ICTY", p. 225.

26 *Accordance With International Law Of The Unilateral Declaration Of Independence In Respect Of Kosovo* was a request for an advisory opinion referred to the International Court of Justice by the UN General Assembly regarding the 2008 unilateral declaration of independence of Kosovo. This was the first case regarding a unilateral declaration of independence to be brought before the court. The court delivered its advisory opinion on 22 July 2010. See more http://www.icj-cij.org/docket/index.php?p1=3&p2=4&code=kos&slu.

27 By a vote of 10 to 4, it declared that "the declaration of independence of the 17 February 2008 did not violate general international law because international law contains no 'prohibition on declarations of independence'"; nor did the declaration of independence violate UN Security Council Resolution 1244, since this did not describe Kosovo's final status, nor had the Security Council reserved for itself the decision on final status.

for rethinking the political *beyond biopolitics*. Such is the articulation of popular sovereignty, as a return to the concept of the people which opens up new possibilities of political and democratic space in the twenty-first century.

But before we turn to the analysis of this issue, let us emphasize once more the call for establishing new international institutions. Even if the appeals for the so-called concert of democracies are not so explicit at present, it appears that such thinking is still at force, especially concerning the role of the UN.[28] The attempts to marginalize the UN, as the international institution *par excellence* – as well as its key constitutive parts such as the Security Council – are perceived in multiple aspects, most significantly precisely in relation to *interventionism* and *humanitarianism*. In this sense, Butler's articulation concerning the war in Iraq had the following emphasis: "the US has decided to suspend its obligations according to the international law. This did not begin with this war. It was the case in Guantanamo, when the Geneva Convention was suspended. Also, the US refused to support the International War Crimes Tribunal, which is maybe the most flagrant way for the US to exclude itself from the international community. US have been suspending its obligations toward international law for some time now."[29]

---

28 In spite of their analysis of politics of "imposing peace", and its role in contemporary biopolitics, Hardt and Negri equally articulate a most extraordinary claim: the concept of the international order is in crisis because in the institutions such as the UN the concept of Empire began to take shape. M. Hardt, A. Negri, *Multitude*, pp. 4–10.

29 Butler, J., "Dehumanization Of The Enemy", in *Filozofski fragmenti (Philosophical Fragments)*, Karpos, Belgrade, 2007, p. 198. Butler argues that a different way of thinking needs to be institutionalized, one which will appear as an integral part of the *ethos* of the community and of US foreign policy. The constitution of such a discourse could begin with a new approach to internationalism. In this respect, Butler also analyzes how the category of life manifested itself in international politics, precisely through the relation of "humanitarianism" against the political. This relates precisely to the issues of de-humanization of the enemy's entire populations. Butler's argument about the paradoxes of "humanitarian interventions" is that they remove and dismiss both the "political" and the "humanitarian", i.e. both the political and ethical aspects claims.

A recent example in this sense is the withdrawal of US, i.e. of both its governmental and non-governmental representatives from the UN international debate "The Role of International Justice in Reconciliation", which took place on April 10 2013.[30] The relevance of this example lies in its peculiar *transparency* and *visibility*: the US, as still the leading figure of the West, refuses a dialogue with "the rest" – or, more precisely, refuses a *dialogue en generale* – on a matter of such great significance for law, justice and politics.

The common signifier in these different examples rests in the reference to *exceptionalism* as the foundation of *interventionist* and *humanitarian* crypto-politics: it opens the space for violence against the other, i.e. against the population of the enemy in a crypto-Schmittian way and as such presents the governmentality of contemporary biopolitics. For, as Heller has articulated, "nowhere is the friend/foe dichotomy so strongly present as in biopolitics."[31] In a similar sense, the example of this dichotomy – and yet another example of it in calls for new institutions – has been perceived by Badiou in his analysis of the European Constitution, particularly in respect to the migration processes, as something Europe needs to protect itself against. Last but not least, the case of Snowden's disclosure of US government surveillance exemplified that a great deal of states are perceived as *potential enemies* and, secondly, that the distinction between the public and the private appears as non-existent in contemporary biopolitical governmentality.[32]

---

30 See more C. Lynch "US To Boycott UNGA Session That Revisits The Balkans Bloody Past", *Foreign Policy*, 10.4.2013. www.foreignpolicy.com.
31 Heller, A. *Biopolitics*, Ashgate Publishing, Limited, 1994. p. 11.
32 Here again we recognize how Foucault's dispositive of security reveals itself in neoliberalism: in the name of US national security information privacy is ruled out. Snowden, on the other hand, wanted to spark a democratic debate on mass surveillance, stating that his "sole motive is to inform the public as to that which is done in their name and that which is done against them". See http://www.theatlantic.com/politics/archive/2014/05/edward-snowdens-other-motive-for-leaking/370068/.

Sovereignty, Democracy And The Political

The analysis of the contemporary phenomena of biopolitics, therefore, discloses that dissolution of sovereignty on multiple levels appears as one of key characteristics of these processes, accompanied by the dissolution of democracy and the political. This happens in reference to the issues of international institutions and international law, but equally concerns numerous aspects of liberal governmenality in public and private life. Sovereignty and democracy, in this sense, relate to the question of political subjectivity, and at the same time to subjectivity and intersubjectivity in *polis* and *oikos*. Moreover, the biopolitical turn here also consists in the unanimity, i.e. in the consensus between the so-called "right" and "left" in the West, in relation to this fragmentation of sovereignty, law and the political, as well as in relation to the replacement of authentic democracy with crypto-Schmittian power politics.[33] Because of this, the rise of discourses on post-sovereignty and liberal democracy – both in opposition to the concept of democracy – manifests different forms of the biopolitical *dictum*.

From Jouvenele to Maritain and Agamben, and backwards, sovereignty appears as absolutized power beyond law and, as such, is presented as always opposed to multitude and pluralism. It is identified with arbitrariness of sovereign's decisions, i.e. with autocracy and untransferability of authority, which testifies to the transcendence of sovereignty as power beyond law. The conceptual lack of these theories, however, is exemplified in the difference between *decesionistic sovereignty* and *popular sovereignty*, the difference they do not address and which deals with the issue of the political.[34] This is the difference between sovereignty as the symbol

---

33 In this sense, the so-called trans-national left progressivists such as A.M. Slaughter, B. de Jouvenele, J. Maritain, G. Agamben, M. Hardt and A. Negri show a high level of agreement with the neoliberal approach in supporting post-sovereign theory.

34 The genealogy of sovereignty in theory is associated with Bodin, while its beginnings in politics are usually related to the Treaty of Westphalia. It is relevant to perceive, as Brierly points out, that neither exercising power in Greek polis, nor Roman *basileus* who has *auctoritas*, nor Medieval doctrines and practices do yet know about sovereignty, i.e. and so sovereignty appears as a *par excellence* manifestation of Modernism. Moreover, the concept of sovereignty,

of absolute power of the sovereign and sovereignty as a symbol of political legitimacy of the rule of the people, i.e. of sovereignty in its structural relation to democracy. Moreover, international institutions such as the UN, especially the role of the General Assembly – in difference to Hardt's and Negri's interpretations – are not examples of decesionism but rather examples of political and legal practices which have as their basis popular sovereignty. The UN is the only place of equality between different states, the only *topos* of *the world assembly* where *each vote counts the same*. In this light, the relation between such a conception of sovereignty and non-interventionism becomes clear, in the sense in which it is defined in Annan's well known article "Two Concepts of Sovereignty".[35]

This approach to the question of popular sovereignty – in spite of the fact that in political theory and political practice it is of later date – has been articulated in the tradition of political philosophy. In such sense, Kalyvas thematizes and argues in favor of a new potentiality of *democratic sovereignty*. In his critique of Agamben, Kalyvas writes that Agamben's conception of sovereignty "does not exhaust the multiple forms and rich democratic potential of sovereign politics" and, furthermore, that "an alternative conception of sovereignty" is possible. Kalyvas also invokes relevant articulations from Plato's dialogue *The Statesman*: presenting Plato's metaphor of weaving, he perceives that it corresponds to "the founding and constitutive activity that makes the web of the state" and, moreover,

---

precisely as a concept of Modernity, manifests the structural ambivalences and potentialities of Modernity – articulating two different possibilities of thinking power and the political. These possibilities are expressed through opposition between authority *vs.* self-legislation, submission *vs.* emancipation and heteronomy *vs.* autonomy. See J. Brierly, *Basis Of Obligation In International Law And Other Papers*, Oxford University Press, 1958, pp. 15–30.

35 In his much discussed article, Annan writes that "in Kosovo a group of states intervened without seeking authority from the United Nations Security Council" and that such precedents are "not satisfactory as a model for the new millennium. Just as we have learnt that the world cannot stand aside when gross and systematic violations of human rights are taking place, we have also learnt that, if it is to enjoy the sustained support of the world's peoples, intervention must be based on legitimate and universal principles." K. Annan, "Two Concepts Of Sovereignty", *The Economist*, 352, September 18, 1999, pp. 49–50.

that "sovereignty can be reinterpreted as an instituting activity that is capable of original political creations and is able to constitute a new political and legal order, but it relies on pre-established materials, activities, and relationships to do so. The sovereign weaver does not start from nothing, and weaving does not represent a radical break with the past. Nor does it have to operate on a social, cultural, and juridical *tabula rasa*."[36]

In short, Kalyvas claims that the potentiality of comprehending sovereignty as creation and action implies that sovereignty, as such, is not a radical movement *ex nihilo* that "begins from nothing", but is an activity which "lies on something". Or, more accurately, it rests on the relation between politics and law, since "the web of politics is also a legal web, the political weaver who ignores the legal threads might fail to create strong and cohesive social bonds. In this sense, the weaver is always already within the realm of the juridical. Contrary to *The Republic*, Plato now considers legality as an important criterion for the categorization of different regime types and for the distinction between good and perverse regimes."[37]

In this sense, Kalyvas articulates the possibility of a new theory of *democratic sovereignty,* in which free speech, critique, public deliberation and reflexive autonomy would have their proper place. Moreover, this movement is one towards a rethinking of democracy different from a form and a procedure, i.e. as a movement which, emphasizing the creative and active participation of the people, lies beyond the so-called proceduralist conception of democracy. In this movement, the political itself comes forth, whereas legality and legitimacy appear simultaneously, through the action of people as equal participants in their common present. Political subjectivity and democracy, therefore, now appear not as exclusive matters of formal structures and procedures, but as exemplary issues of intersubjective creativity and production. In such a discourse, law, popular sovereignty and democracy – in the sense in which they constitute the political – are opposed precisely to biopolitics.

---

36 A. Kalyvas, *Politics, Metaphysics And Death*, ed. A. Norris, Duke University Press, 2005, pp. 118–127.
37 *Ibid.*, p. 127.

Similar to Kalyvas's argument regarding the structural relation between sovereignty and democracy, Manning writes that "democracy is usually understood as a political regime and a system of institutions that enable popular sovereignty."[38] Moreover, in respect to the relation between democracy and freedom, and in reference to the political, in her analysis of contemporary biopolitics, Heller emphasizes that "today democracy has been widely separated from its twin term mainly for two reasons: the rise of economy as a public institution substituting older concepts of *res publica*, defined as genuinely political, and the erosion of the nation as the traditional framework of the body politic through the trans-nationalization of economic transactions."[39] Moreover, considering the relations between democracy, law and freedom, Heller articulates that "biopolitics misunderstands democracy", and that precisely because of its multiplicity, "democracy presupposes and aims at freedom and equality of all *citoyens* and *citoyennes* participating in a political community. Democracy is based on openness: if one particular group requires the closure of debate in the name of a specific interest, democracy comes to an end and totalitarianism enters the scene".[40]

Indisputably, in the discussion of what constitutes a legitimate civil order of popular sovereignty and freedom as its constitutive part, we can recall Rousseau's *The Social Contract*, i.e. the idea of the people that it expresses, as "a form of association that defends and protects the person and goods of each associate with the full common force, and by means of which each, uniting with all, nevertheless obeys only himself and remains as free as before."[41] Precisely in this sense, Rousseau differentiates regulative reason from calculative reason, emphasizing that in the social contract, as creation of popular sovereignty, an ethical transformation, in which each conceives oneself as a member of political community, takes place as well. Rousseau, finally, draws the line between popular sovereignty and decisionistic sovereignty by arguing that "if a people promises simply to obey, it dissolves itself by this very act, it loses its quality of being

---

38 E. Manning, "Time For Politics", in *Sovereign Lives*, Routledge, New York, 2004. p. 74.

39 A. Heller, *Biopolitics*, p. X.

40 *Ibid.*, p. XI.

41 J.J. Rousseau, *The Social Contract*, Cosimo Inc., New York, 2008, p. 49.

a people; as soon as there is a master, there is no more sovereign, and the body politic is destroyed forthwith."[42]

It is in the popular sovereignty that constitution of the rule of law appears as tied with the democratic principle, referring to the sphere of autonomy and freedom, where sovereignty is presented as a symbol of political legitimacy. Furthermore, the political itself therefore appears as the democratic – where there is no space for democracy one cannot speak about the political either. Or, in Arendtian terms, if democracy presupposes the existence of a body politic, in contemporary politics it is becoming less and less clear by whom this body is constituted and who sets the rules. Consequently, this is why the biopolitical discourse about the end of history, as one of the leading discourses of contemporary neoliberalism, has simultaneously been followed by the discourse about the end of democracy as well, as Guehenno has shown.[43] Guehenno has also claimed that conceptions of world parliament and world state will always remain utopian, and that the end of institutions in their existing forms will be the end of the political as well.

Moreover, in this sense, in biopolitical governmentality, norms and rules become regulations of economic transactions which do not have to be articulated and implemented by political and democratically legitimized actors, but simply by the so-called experts. Precisely in such a world the citizen as the constitutive element of a sovereign political community becomes obsolete. In opposition to such discourses and practices, throughout the tradition of political philosophy from Plato to Rousseau and Kalyvas – the relation between the political and the legal, i.e. between politics and law rises as properly constitutive for the political. In this sense, politics of democratic sovereignty as politics of creative sovereignty simultaneously exemplifies a multiplicity of traditions, a set of localities and specificities of potentials for sovereignty. Furthermore, this is one of exemplary and rare issues in which politics – as established on creative sovereignty – arises both as a field of possibilities and as a field of preservence. In addition, not only is it the case that these two separate features do not need to appear as mutually excluding, but the political itself presents them as its constitutive

---

42 *Ibid.*, p. 57.
43 J.M. Guehenno, *La fin de la democratie*, Flammarion, Paris, 1999.

elements, from which the relation between politics and ethics comes to be established.

In her analysis "Has Biopolitics Changed the Concept of the Political?" Heller examines whether biopolitics can be called politics at any cost, and argues that from the era of the Greeks and the Romans up to the entire European tradition, politics proper has meant in the most important sense *denaturalization*. Recalling Hegel's discussion of self-consciousness in the *Phenomenology of Spirit*, Heller articulates that politics proper relates to "the common thing":

"One could say without grave exaggeration that biopolitics is a legitimate heir of nineteenth century radicalism, of the reversal of Platonism, to use Heideggerian jargon. Platonists in reverse are seeking to establish the non-spiritual essence of all human manifestations, this can be the economic, the biological, the sexual, the instinctual in general and the like".[44] Tracing biopolitics in Arendt's study *The Origins of Totalitarianism*, Heller considers this discourse in the discussion about modern racism, emphasizing its self-contradictory character, in the sense in which politics always relates to the appearance of something new, i.e. to the creation of and bringing into existence a yet unseen potentiality.

Biopolitical thinking, conceptualized either in terms of race or sex, or in any other form, does bring about the birth of the new in the spheres of visibility, nor does it so in public and political dialogue, in the form of active participation of multiplicities. Moreover, Heller articulates how such biopolitical discourses have established themselves as philosophy and science: "if there are university departments of gender studies, gender history and gender literature, the whole biopolitical movement of gender has legitimated itself scientifically."[45]

Ojakangas has also perceived that "power of biopolitical societies is not political power at all, but purely administrative power – power of experts and interpreters of life".[46] Moreover, such "interpreters of life" must be logically superior by virtue of their knowledge of the processes of

---

44 Heller, A., *Biopolitics*, p. 5.
45 *Ibid.*, p. 9.
46 M. Ojakangas, "The End Of Biopower", *Foucault Studies* 2, 47–53. 2005, p. 16.

life which they aim at securing, and this is how contemporary biopolitics manifests itself as a movement against democracy. In response to this situation, Heller argues for a "framework of a new critical theory", which would equally address the issue of the individual being thrown back into itself, since this has produced consequences such as a retreat to ethics in contemporary critical thinking (as in postmodernism, for instance). Moreover, in this context, Heller also argues that new political theory cannot remain exclusively normative, i.e. that it also must conceptualize itself as a critical theory which seriously examines contemporary biopolitics, which is occupying the place of the political.

Finally, Foucault's claim that the first issue of philosophy is the question of the present, the question who we are today, is situated in this discussion as a response to biopolitics. The key to comprehending the bottom line of biopolitics lies in conceptualizing that "politics is immanent to history and history is necessary for politics", and that both, as such, relate to philosophy. This means that different phenomena should be analyzed as they appear in the given historical context, in their relation to democracy, sovereignty and freedom – as relations *beyond biopolitics*. If "one of the decisive characteristics of Western societies is that relations of forces that found their manifestation for a long time in war", and that "step by step these relations entered the order of political power",[47] as Foucault argued, then the perception of a genealogical-critical discourse, and its conception of freedom, is a response to the doctrine of liberal interventionism on all levels. In this movement, the political itself appears in such a relation between freedom and democracy, as a form of resistance to multiplicities of contemporary biopolitics. And unlike Agamben's, Hardt's, Negri's and Derrida's assumptions, this freedom is not one of a messianic expectation. In its structural relation to the political, it is the prerequisite of the possibility of democratic sovereignty as popular sovereignty. The response to biopolitics, therefore, lies precisely in the relation between sovereignty and what Foucault called "revolutionary discourse", as a yet unexplored possibility of political philosophy and democratic practice. Moreover, one could in this context, in Foucauldian terms, rethink the bond between "the scientific

---

47 M. Foucault, *History Of Sexuality I*, Knopf Doubleday Publishing Group, 1990, p. 116.

knowledge" and "the knowledge of the people", or between "the scientific knowledge" and "local memories". In such a project of liberalization of historical knowledge – in opposition to scientific formalism – the political finds its place in the opening of new potentialities, i.e. in the multiplicities of new political subjectivities in popular sovereignty.

Furthermore, it is worthwhile noting that Foucault's line of differentiation between "the revolutionary course" and "the utilitarian course" relates to the gap between *law* and *practices*. In the framework of phenomena of contemporary biopolitics this rupture appears as the difference between popular sovereignty and international law *vs.* liberal interventionism and its practices. Finally, in the initial political acts, such as refusal, critique, imagination and creation, which are, according to Foucault, simultaneously "political, ethical, social and philosophical", a new relation is established between the critical and the normative. At the same time, this is a task where the issue "who are we in a time that is ours" is addressed.[48] "The task today is not to discover who we are, but to *refuse what we are. We have to imagine and invent what we could be…We have to promote new forms of subjectivity*, refusing the type of individuality that has been imposed on us…"[49] Here *refusal* appears as the exemplary *political* and *existential* act, as the refusal of what we are in *contrast to what we could be*. This implies that the political act is represented as an act of *imagination*, a reopening of the field of *possibilities* and human *creation*, which opens the doors for *invention, political action* and *realization of such potentialities*.

---

48 In this sense, Foucault writes that the leading idea is to "to give oneself the rules of law, the techniques of management, and also the ethics, the ethos, which would allow these games of power to be played with a minimum of domination." 284 284 Foucault, M. "The Ethics of Care", p. 18. Moreover, it is in this sense now that Foucault's theory arises as deeply and profoundly affirmative, and not "negative" or "nihilistic" (Fraser 1995, Walzer 1986), and in contrast to every interpretation that attempts to present Foucault's freedom as a "private creativity of the individual" (Rorty 1992).

49 Michel Foucault, "Subjectivity And Power", in H. Dreyfus, A. Rabinow, *Michel Foucault: Beyond Structuralism And Hermeneutics*, University of Chicago Press, Chicago, 1993. p. 209.

The act of refusal, followed by the creation of the political, between Lefort's *la politique* and *le politique*, makes it possible to articulate political subjectivity on multiple levels. For, "if to politicize means to return to standard choices, then it is not worth it. To new power techniques one must oppose *new forms of politization*".[50] Such thinking discloses the way for the birth of democracy as creative sovereignty, as expressed on different levels of theory and practice of political subjectivities. This creative sovereignty appears as the opening of the gates for freedom, and it becomes in multiple forms of pluralities, communities and intersubjectivities a living and authentic form of democracy. Moreover, the political itself, in this sense, appears as a response to events but equally as the initial act both of creation of events and creation of democracy. Democracy, therefore, besides contingency and proceduralism, is significantly characterized by its relation to the possibilities of political action, and this is where genealogy of the present finds its important place. Democracy of local popular sovereignties, therefore, is the event of the constitutive and the constituting of political subjectivities; it is a response to biopolitics. Moreover, the idea of the *many*, of *multiplicities*, is a prerequisite of democracy and, in this sense, of the political as well. If a *single name*, therefore, is articulated so as to comprehend wholeness, then this act is by itself a movement against democracy, since democracy is one of the multiplicities of political subjectivities. Similarly, the other prerequisites of the possibilities of the political refer to the relation between popular sovereignty, law and freedom, and potentialities for constituting different forms of political multiplicities. In response to contemporary phenomena of biopolitics, conceptual and empirical possibilities manifest themselves in the interplay between creative sovereignty and democracy.

---

50 M. Foucault, "Power Affects The Body", in S. Lotringer *Foucault Live: Interviews, 1961–84,* Semiotext(e), 1996. p. 209.

II

Twenty-First Century – Time Of Small Revolutions?

Twenty-First Century As The World of Biopolitics

In previous investigations we have reflected on some relevant phenomena of contemporary biopolitics – such as international terrorism, bioterrorism, new international institutions, neoliberal "humanitarian" interventionism, the question of international tribunals and the blurring between the public and the private. However, in spite the insights coming from such examinations, it still has not appeared as plausible that the world of the twenty-first century arises as the world of biopolitics *par excellence* i.e. that *biopolitics* is the name of the time that is ours. This means that, in a yet unseen change in the flow of time in the era of (post)globalization, in the theoretical, but even more in the political sense, a specific turn has manifested itself and, moreover, one in which new shapes of biopolitical concepts and practices unfold and multiplicate themselves. These newborn processes are reflected in the world economic crisis as *the crisis of neoliberalism per se* and they are demonstrated in the growth of *class differentiation,* as well as in many attempts to *control and govern entire populations.* They are equally visible in a series of *new wars and interventions*, in the concentration of forces of different "*friends*" and "*enemies*" and, not less significantly, in the growth of *simulation of politics* in many governmental and discursive mechanisms and in the *weakening of democratic capacity* for key Western projects such as the *EU.* On the other hand, practically simultaneously with this dominance of biopolitics, completely opposite events, which testify to a silent but incisive *return of the political,* have appeared as well.

The situation has become even more complex since contemporary societies are influenced by the fact that only in rare *clara et distincta* cases it is possible to draw a clear line of differentiation between the radicalization of political and discursive processes which undermine the political, and the forms of exemplary political practices which are directed towards structural transformations. Whereas the economic crisis as systemic crisis of neoliberalism (*ergo,* the crisis of forms of governmentality which Foucault has termed biopolitics) has clearly outlined the limits of a world which corresponds to the domination of the West in political, financial, military, scientific, cultural and technological sense,[51] it has also produced even stronger unification of the representatives of the *status quo* i.e., their joining into a single front of struggle for survival and power. *Crypto-elites* of large capital, in new, and perhaps in the last phases of their dominance of biopolitics, have found themselves practically face to face with humiliated poor *people* against which their economic, political, social and cultural activities are directed. Moreover, the struggle for *bare life* and *dignified life* has become *one and the same*, and the quest for survival and humanity are clearly counterposed to the idea of progress initiated by crypto-elites.

In other words, the level of proximity among these two opposed groups – *the people* and *crypto-elites* – is on the rise, and so is the perception of the exploitation of the majority of population as the condition of survival of the few who govern. On the other hand, and precisely because such exploitation has become almost *total,* reaching over all spheres of human existence, the re-collection of the people has began to – sporadically or more frequently, spontaneously or in a more organized fashion, on the level of incidents or on the level of massive protests – to encompass the public space and horizontally strengthen, seeking the way for *forms of new politics.*

In such time, in which the lack of conceptual articulation is becoming obvious, and in which the space for authentic political action is still relatively small, *imperial crypto-elites* are using all disposable resources for establishing the category of *Schein* in divergent aspects of quasi-economic, quasi-political and quasi-scientific *hyper-production of reality*[52] or – in

---

51 From the standpoint of foreign policy analyses, it is necessary, in this sense, to examine all theoretical and practical causes and effects not only of the emerging world significance of countries which belong to BRICS (Brazil-Russia-India-China-South Africa), but equally of the so-called "next eleven" i.e., Bangladesh, Egypt, Indonesia, Iran, Mexico, Nigeria, Pakistan, Philippines, Turkey, South Korea and Vietnam. There are, therefore, reasons to presuppose that – beginning from 2014 – in next decades multiplication of non-Western discourses will take place and, moreover, that these discourses will condition and be conditioned by new political and social practices.

52 In Hegel's Logic *Schein* is used to refer to appearance, but such appearance which – unlike *Erscheinung* – points to the *hiding of essence*. In the Marxist approaches this conceptual distinction continues and, moreover, expands, in its of articulation of what philosophy of *praxis* is opposed to.

Baudlliard's terms – *production of hyper-reality*, of various *simulacra* and *simulations*, that leads to creation of *virtual worlds*. So, for example, the production of new realities in the situation of economic crisis is accompanied with interpretations that suggest the change of (neoliberal) system is not necessary i.e., that it can be corrected by adopting and applying a series of measures (the most obvious example being the insistence on "austerity measures" in spite their effects). The rule of *virtual worlds* is the rule of inversion: what *appears* as true is, in fact, not true, while *what is true does not appear* but remains concealed and practically invisible.

Moreover, the *simulation* present in such new discourses refers to the building of a strong illusion that, say, so-called austerity measures are sufficient to save the system and, moreover, propose this now as the desired aim of our *"new reality"*. Simultaneously, the process of founding new *political virtualities* is concentrated in three ways:

(1) in the continuation of a specific *mimesis* of authentic political elite (in this sense, states in which those who govern actually represent the will of the people are exceptions);

(2) in the expansion of new *micro-imperialisms* under the parole "better life for all";

(3) in the continuity of "democratic interventions" as "interventions for freedom" (this aspect relates to both "internal" and "external" interventions).

On closer examination of the first aspect, i.e. the issue of *mimesis of the political,* one should, first, comprehend the continuity of the *simulation of the politics of left and right* which comes forth from the *consensus of crypto-elites (active consensus of crypto-elites as opposed to passive non-consensus of citizens)*.[53] For, the *consensus around mimesis,* around the simulation of politics with the aim of gaining even more power, has opened the space for the final erasure of the key political differences between left

---

53 *Politics of left* cannot exist if its core questions are not *justice, equality* and *freedom*, i.e. it is by its concept opposed to every form of imperialism and domination, while *politics of right* cannot exist if it does not deal with issues of *national identity* and *sovereignty*, and the relation between *the national moment* and *the democratic moment.*

and right and, consequently, excluding even the possibility of appearance of political emancipation and/or political self-consciousness.

Ultimately, in this way it is secured that even the most dramatic implications such as, in particular cases, destruction of entire states, *goes practically unnoticed*, i.e. that in the beginning of the twenty-first it is not yet understood to what extent the marriage between postmodernity and neoliberalism has relativized not simply the idea of sovereignty but the whole content of classic concepts such as *territory* and *people*. The relativization of these concepts, as well as the imposition of new meanings which are supposed to affirm that there is nothing exceptional in the fact that most relevant decisions frequently occur far from the eyes of the public – or that usually it is anything but clear who are the real subjects in power – builds a thread between two significantly different theories of biopolitics, i.e. *biopolitics as neoliberalism* (Foucault) and *biopolitics as permanency of "state of exception"* (Agamben). Or, founding a new "practice of truth" as *the practice of reality*, which significantly appears as *consensus around mimesis*, has conditioned the weakening of law and sovereignty, affirming new processes of fragmentation and dissolution of the political to such an extent that, in first decades of this century, contemporary biopolitics manifests itself through series of actions which refer to disappearance of *political subjectivities per se*.

But let us now turn to the second aspect of the power of *political virtualities*, i.e. to parallel expansion and growth of *new micro-imperialisms*, and its parole of creating a "better life" for all citizens. In such sense, for example, the constituted gap between the "South" and the "North" of Europe actually appears as essential strengthening of the central position of Germany, demonstrated in its economic, political and cultural superiority and domination over "poor southern people". Here it is necessary to keep in mind that, simultaneously with this type of real-political establishment of *micro-imperialisms*, official public discourses still insist on the postmodern-neoliberal concept, i.e. that no European state should be for itself or, more precisely, that each has to be in favor of a broader unit – here EU structures – so that *"better life"* can be secured.[54] In this sense,

---

54 The case of Greece, where the government exists only to serve the interests of international financial institutions and the EU, is exemplary: any possible

the so-called superiority of structures is exemplified as the path for the overcoming of the crisis of entire populations of the unemployed, the "excluded", all *sans papiers* or *banlieues,* while, at the same time, the control mechanisms over states and populations are on the rise.

Finally, the third aspect of contemporary biopolitics refers to the growth of violence and criminalization, strengthened by the founding of a previously described system, and which point to an ever-present possibility of intervention regarding domestic political and economic issues of different states. This further means that the neoliberal doctrine, in new circumstances, has developed subtle variations of the already present overturning of Clausewitz's formula articulated by Foucault, i.e. the situation in which *war becomes the code for peace* or violence appears as the other name for peace.[55] Moreover, we can say that, through simultaneous dissolution of sovereignty and law, and the rise of multiple types of violence both in the public and the private sphere, we are witnessing a special encounter of Foucault's and Agamben's theory on the level of biopolitical practices, because "bare life" now establishes itself as the primary category for the majority of population and as the sign of the definite erasure of the difference between *logos, praxis* and *brutal violence.*

In such light, philosophy should articulate the *sine qua non* condition of these three aspects of the production of political virtualities: the disappearance of great intellectual debates or a complete marginalization of what is left of them, i.e. the realization of "the death of the intellectual" which also means the exclusion of the possibility of their new appearance. For, the project of establishing the *hyper-media world of postmodern production of the real*[56] is conceptualized to appear as the *end of (critical) thinking,* on

---

claim of Greece to be a democracy has been rendered absolutely invalid. In such a situation, Tsipras and Syriza are sending the message that anti-European impulses are precisely one of the EU. "We are against casino-capitalism and the imperative that profit comes before the people", says Tsipras, noting that "great and substantial changes in global system of governing are underway". See http://greekleftreview.wordpress.com/category/politics/page/32/.

55 In this sense, indisputably, it is ironic that in the same year in which its member states have, explicitly or implicitly, intervened in a series of operations in Africa, EU has been awarded the Nobel peace prize.

56 Doubtlessly, the society which Debor has described as the society of the spectacle expanded in the contemporary media dominated world. However, the role

the imaginary horizon of the present and mythologized future as exclusive time dimensions, and in the *topos* of *megalopolis* (in contrast to *cities,* i.e. to *polis*). In this sense, the expansion of the so-called *knowledge of the experts* has marked the first step towards gradual disappearance of the power of the citizen as *zoon politikon,* but also a step towards "superseding" the figure of *the intellectual in public discourse.* Finally, this process was composed to lead to total passiveness which would make the idea of freedom senseless, unimaginable and unnecessary. For the simple reason that the presupposed "triumph of neoliberalism, through synchronized sacralization of "better life", has proved to be a triumph of the virtual, still immanently biopolitical world, which signifies governmentality over the entire population.

However, the situation which can be perceived as a *par excellence biopolitical situation*, and which as such appears as a key example of the radical gap between *forms* and *reality* – as a rupture between *anti-politics* and *democracy* – relates to the issue of expelling *popular sovereignty* from political practices and public discourses.[57] On such basis, the production of *quasi-polis,* which also includes the wholeness of *bios,* has become plausible, while at the same time this totalizing process, expanding to spheres of art, culture and science, has enabled nihilistic relativization of all values. And in the world in which there are no more values worth living for, i.e. in which the exclusive value is presented as *"bare life"*[58] and *survival* – thinking becomes *completely needless.*

---

of media is certainly not one-sided, as the examples of different uses of social networks have shown. Or, the example of Glenn Greenwald's journalism in the Snowden case, can be perceived as an example of media contribution in the struggle for the new appropriation of the public sphere.

57 In *The Democratic Paradox* Chantal Mouffe has noticed that expelling popular sovereignty from theoretical, political and public discourse has been a strategic mistake even from the perspective of neoliberalism and moreover, a mistake which consequently contributed to the growth of democratic deficit. Mouffe also claimed that "this can have very dangerous effects on the allegiance to democratic institutions." See C. Mouffe, *The Democratic Paradox,* Verso, London – New York, 2005, p. 4.

58 On the concept of *"bare life"* (*nuda vita*) see G. Agamben, *Homo Sacer,* Stanford University Press, Stanford, 1998.

Arab Spring Between Biopolitics And Democratic Uprising Of The People

How are we now to interpret the phenomena of the so-called Arab Spring, i.e. a series of turbulent uprisings throughout the Arab world, which have not only structurally shaken the political order, producing political earthquakes of extraordinary dimensions, but have also appeared as theoretical and practical symbols of the birth of a new world, as *events* which surpass local frameworks and potentially expand beyond all existing borders? Are we faced with authentic opposition to multiple forms of domestic and foreign imperial politics, with the democratic uprisings of the people against totalitarian regimes? Or is something very different at stake – a *biopolitical process* which actually suggests the continuity and strengthening of the neoliberal trans-national Atlantic order in Africa? And how far is justified the thesis that the original becoming of *political subjectivities* is here taking place, i.e. that a form of "Arab Enlightenment" has reached its peak, bringing forth a particular political self-consciousness and emancipation? And how much truth lies in the insistence that this is an exemplary case of contemporary *political virtualities,* of the simulation and production of a *quasi*-democratic revolution, whereas, in reality, we witness strong influence of mostly particular Western interests in these areas? What are the causes and what are the effects of political and cultural, essential and existential, conceptual and material transformations of the Arab world which are taking place in front of our eyes?

In *The Rebirth of History: Times of Riots and Uprisings*[59], Badiou articulates a typology of uprisings, in which he particularly turns to events in Tunisia and Egypt, and this is perhaps the proper place to begin our analysis. Badiou's discussion of concepts of *"change"* and *"reform"* which – while proclaiming "modernization", "democracy", "international community", "human rights" and "globalization" – in reality fix "the value system of imperialism", i.e. of *contemporary barbarism*, offers a path for our approach

---

59 This is the only philosophical work which extensively deals with the Arab Spring. See A. Badiou, *The Rebirth Of History: Times Of Riots And Uprisings*, Verso, London – New York, 2012.

to the Arab Spring. In the name of such "reforms", as Badiou writes, bloody military expeditions, especially in Africa, are undertaken, so that the respect of "human rights" should be secured. This so-called respect for "human rights" refers to the right of the powerful to divide states or to bring to power, through combination of violent occupation and phantom "elections", their corrupt servants who will place all relevant resources of Arab countries at the disposal of the powerful for nothing.[60]

On the other hand, Badiou is simultaneously persuaded that the present, as a whole, brings forward first traces of *global popular uprising* as a rising against such regression i.e., as an opening of *time of riots* in which the *rebirth of history* is taking place, adding, moreover, to this, his well-known programmatic thesis that this equally must be a *rebirth of the left as well*.[61] It is important to keep in mind that, for Badiou, unlike for Hardt and Negri, *the rebirth of history* in no way follows by itself from contemporary capitalism and the actions of its protagonists, but rather implies the birth of both destructive and creative capacities, aiming to bring to an end the existing order.

In his analysis of different forms of *immediate riots*, Badiou points, firstly, to the riots of young and poor people in the suburbs of European cities, connected with the clashes which occur in the dark areas of our megalopolises, recognizing that it is even permissible to kill the so-called *banlieues*, because the death of such people means nothing to those in power. For us, this is of extraordinary relevance, since it refers to *contemporary biopolitical phenomena per se*: while there is infinite tolerance for the criminal acts of bankers, as paradigmatic representatives of contemporary neoliberalism, there is, at the same time, zero tolerance for petty criminal acts of the poor, even a silent permission for their murder. Moreover, the *mainstream* discourse which is prone to point to "dangerous classes", insists that they are simply criminals – and precisely such *criminalization of the enemy*, i.e. naming the enemies of the system as *criminals*, appears as the first thesis which we, in our analysis of the Arab Spring, attempt to

---

60 *Ibid.*, p. 4.

61 In a similar fashion, Badiou articulates that in contemporary capitalism we can no longer speak of differences between the so-called left (Obama, Zapatero, Papandreou) *vs.* right (Sarkozy, Merkel) governments. See *op. cit.*, p. 14.

articulate. Is it not true that this phenomenon illustrated the exemplary situation of contemporary biopolitics? Moreover, it also exemplifies the very foundation of biopolitical systems of control and governing which is applied in a series of different practices, in the wide specter of foreign and domestic politics in numerous Western centers of power.

Is not this type of legitimizing the right to kill structurally similar to the so-called right to intervene, which is used as the explanation for interfering with domestic issues of sovereign states, while, at the same time, criminalization of their leaders is taking place? Because, *biopolitical discourse* states that *banlieus are hooligans,* and state leaders are selectively and on arbitrary basis named as *totalitarian dictators*, while in both cases we are dealing with extraordinary forms of *criminalization* with the aim to convince the public that practically every activity, including assassination, is completely justified and even desirable. We should also note that the first task of these biopolitical discourses is the establishment of general consensus, precisely in the sense in which Ranciere emphasizes that the greatest contemporary crime is life in "the *time of consensus*", which causes unquestionability, dogmatism, erasure of critical thinking and practically makes *invisible* and *inexistent* all those *excluded* from the *consensus*.[62]

Another striking example of the consensus of imperialistic discourse, i.e. of the erasure of the difference between right and left, and of biopolitical practices *per se*, is the existence of secret lists for assassination in the US,[63] as something over which a broad consensus between different political alternatives is established. It is equally relevant to pay attention to the so-called classification of the innocent victims in US attacks by unmanned aerial vehicles (UAV), i.e. to the fact that all men who find themselves on the place of attacks are marked as "militants" that is, as "US enemies". Biopolitical turn of such a situation – relevant for our further analysis – becomes entirely transparent in the demonstration of power in deciding who can and who cannot be killed or, more precisely, assassinated without proper trial and without any reference to either legal or ethical norms. This is structurally similar to naming someone a criminal without any evidence

---

62 J. Ranciere, *Chronicles Of Consensual Times,* Continuum, New York, 2010.
63 J. Becker and S. Shane, "Secret Kill List Proves A Test Of Obama's Principles And Will", *The New York Times*, May 29 2012.

for the crime committed, but it is also accompanied with the "theoretical-political" implication that in reality *anybody* and *all* are *potentially suspect, ergo,* practically anybody is a *potential candidate for the enemy.* The radical and new moment here lies in the fact that contemporary neoliberalism, i.e. biopolitics, has produced the condition in which the mechanism of usurpation of power and its use for brutal violence turn against their own citizens as well – for now they also appear as legitimate targets of attacks, and consequently what becomes questionable is not "only" the international legal order but domestic legal foundations of the state as well.

But let us now return, in our discussion of the Arab Spring, to Badiou's typology of riots and uprisings. Articulating the difference between *an immediate riot*, *a latent riot* and *a historical riot*, Badiou writes that the uprising in Tunisia, which represented the beginning of a series of "Arab revolutions" in 2011, was primarily an example of an immediate riot, which most frequently happens as the initial form of a historical riot. Emphasizing that the participants in immediate riots are mostly young people, Badiou refers to the examples of 1966–1967 China, as well as 1968 France, both of which testify to the key role of students. The secondary feature of this kind of riot is its localization, i.e. the limited territory where it takes place. *The localization of a riot* is of fundamental importance, because it is only when a riot spreads from suburbs to *the city centre* that it transforms its shape and becomes a *historical riot.* In this light, an immediate riot, which is limited to its own social space, cannot by itself appear as a force of *powerful subjective political transformation,* for it dissolves in the destruction of the symbols of contemporary neoliberalism. It is only when an immediate riot spreads to population which is, either by its status, social background, gender or age, far from its constitutive center, that we can speak about a *historical riot.*[64] This is why immediate riots have neither political, nor even pre-political character – at best, they open a space for *historical riots,* presenting traces of the fact that the system is not invulnerable and that authentic ruptures exist.

---

64 A. Badiou, *The Rebirth Of History: Times Of Riots And Uprisings*, pp. 24–25. Since Badiou's idea of latent riots does not touch on Arab Spring, but is relevant for the issue of contemporary revolutions, we will turn to it in our last chapter.

What is relevant here for us is the following: Badiou believes that we have a lot to learn from the Arab countries about *historical uprisings,* and the transformation of an immediate riot into a historical riot. Discussing *sine qua non* conditions of such an event, i.e. the transition from limited localization to the central place, the essential peaceful positing of rebels in such a place and the extension of time which accompanies this (because, unlike an immediate riot a historical uprising can last for weeks or months), Badiou articulates what we can call moments of *new authentic happening of the political* or the *event of the political.* Apart from the relevance of *place and time,* i.e. the occupation of a *central place* (which becomes *historical)* and the extension of time, the decisive significance lies in *the beginning of subjectification.* This means that what is at stake is the so-called "qualitative extension" in which different parts of population gather and collect – the young, the working class, the intellectuals, families, women, state employees, even police officers and soldiers, building a *multiplicity of voices* which begins to call itself *the people.* Finally, a particular slogan, that articulates the essence of uprising, is created, and it is as if that which the riot is about has *already been decided.* For Badiou, who particularly insists on examples of Tunisia and Egypt as representative examples, this is sufficient to say that *a rebirth of history* is taking place, and that we are dealing with the opening of the horizon for the appearance of a new idea and political emancipation.

But this is a place to pause and rethink. The immanent difficulties which arise for Badiou's theory bring us simultaneously *in medias res* of political-philosophical considerations of the Arab Spring. Although, on the one hand, it is indisputable that the majority of these "revolutions" inherently and structurally contain in themselves something of a *"democratic moment",* it is difficult to name them as exclusive carriers of political subjectivities. In other words, in spite the potential return of the political present here – which testifies that we are not dealing with a clear case of biopolitics as in the previous examples of criminalization of foreign and domestic "enemies" – the implications are still twofold and ambivalent. The first argument why this is so lies in the *insufficiency of the idea*: for, to speak seriously about a *historical uprising*, especially an uprising which should turn *a new idea into reality*, it is not enough that the slogans of the protest should be "Ben Ali, go!" or "Mubarak, leave!". Certainly, aware of this difficulty, Badiou notes that we do not really know what the "historical uprisings" in Tunisia, Egypt, Syria and other Arab countries are going to lead to, suggesting that this is the *time of intervals* which can realize different historical possibilities, i.e. that we are speaking only about the possibility of the appearance of a new idea, and that history in itself does not contain a solution for the problems it places on the agenda.[65]

Now, these observations are doubtlessly correct, but they do not relate to the described problem because nobody claims that a *historical uprising* needs to provide *clear solutions*. But, *to be historical* – and this is in high proximity to its *authenticity*, its *political* and its *historical* consequences – an uprising certainly has to contain a clear *trace of an idea or ideology*, and not merely of its *possibility*, since in an uprising practically all, completely different possibilities, appear, and most often a set of contingent circumstances makes one or the other prevail. Moreover, neither of these "possibilities" are in opposition to the so-called "revolution", since, aside from resisting *the status quo*, they *have not said anything about anything else.*

We could develop this and take it further in many ways, all of which are practically relevant, but essentially this idea implies that a *historical uprising* is another name for a *revolution*.[66] But apart from including multiple parts of a population, occupying a central space, extending time and offering its slogans *against*[67], it would have to contain an undivided "*yes!*" so that it could be properly called a *political and subjectifying event*.[68]

---

65 A. Badiou, *The Rebirth Of History*, p. 42.
66 There is a conceptual difference here i.e. revolution presupposes that resources for taking over power are already present, while historical uprising does not. This difference is of little relevance here.
67 In this light, even the slogan "Al-Sha'b yurid isquat al-nizam!" ("People want the fall of the regime!") appears as a slogan *against*.
68 From this perspective it becomes clearer why the so-called October 5 2001 in Serbia *has not been a historical uprising of the people i.e. neither a revolution nor an event*. While, on the one hand, there existed a significant *against* the lack of a clear *for* which would refer to a shared idea or ideology, showed that one cannot speak here about *the political, about emancipation and about birth of a subjectivity*. Moreover, the split, even strong opposition in persuasions between key political protagonists, as well as lack of a clear political bond, consequently has been the cause of political and social chaos for in next decade.

Moreover – and this is a relevant difference from a *contemporary phenomenon* we will analyze and which happens in *the West* – this is even more an imperative when we consider the cases of the *Arab countries*. Because – and this is a second problem for Badiou's theory – in the context of contemporary neoliberalism, the influence of foreign interests and their imperial discourses and activities in these "spontaneous uprisings" cannot be underestimated. This is particularly the case if we keep in mind that in these areas key Western countries do have their interests, and that sometimes, explicitly or implicitly, the *"right to intervene"* has been used a great deal.

This places the entire issue of the Arab Spring to the level of biopolitics, particularly if we recall the produced hysteria and *criminalization* of overthrown leaders in Arab states. But everything in the in the so-called "springs" cannot be reduced to biopolitical issues, i.e. the issues of staged and directed events which result from control and governmental mechanisms, as a crime of simulation of new political virtualities. Seemingly paradoxically and ironically, in favor of such an interpretation goes the fact that the *time of intervals* will, all the odds are, articulate ideologies opposed to those the Empire thought would be of use. For, Western interpretation of the Arab Spring – as the interpretation which insists that Arabs fight for freedom, meaning by this capitalistic "freedom of thought" and "free elections" between actors that present *the other of the same* – is attempting to say that Arabs *desire the West,* that they desire the *pleasure* of the Westerns and dream to be a part of that "civilized world", but the West overlooks that this is its own *produced interpretation*.

In this respect, perhaps the most striking example is the public debate between Alain Badiou and Jean Luc-Nancy concerning the war in Libya. While Nancy's article indisputably presents an open support for Western intervention, it also, at the same time, re-affirms, in a new way, the biopolitical discourse about "the role that the West should play". Or, more precisely, and as a paradigm of previously noted production of interpretations and desire for the West, Luc-Nancy articulates the basis of a *neo-colonial*

As a result, the state was reduced to neoliberal, *ergo*, biopolitical domination, where the level of control exemplifies, as in numerous states, that we speak about *neo-colonialism per se*.

discourse as its protagonist. *"It is not that the poor old 'West' has cleaned up its act"* because, simply, "in the process of melting in the fusion that begets another world... it is necessary to reinvent the act of living", and *"it is necessary to strike... It is a delicate task. But at stake is what we want to live and how we want to live it, with an acuteness that we are not accustomed to. That is what the Arab peoples are also signifying to us."*[69]

Badiou's response to such a discourse has been the following: "Didn't you notice right from the start the palpable difference between what is happening in Libya and what is happening elsewhere? How in both Tunisia and Egypt we really did see massive popular gatherings, whereas in Libya there is nothing of the kind? Didn't you know that the French and British secret services have been organizing the fall of Gaddafi since last autumn? How can you of all people fall into this trap? Can you simply accept the 'humanitarian' umbrella, the obscene blackmailing in the name of victims? Do you believe, can you believe, that they represent 'civilization', that their monstrous armies can be armies of justice? I ask myself what good is philosophy if it is not immediately the radical critique of this kind of unreflecting opinion..."[70]

It is, therefore, a separate issue how *specific* are uprisings in different Arab countries, i.e. are the particularities so relevant that they transform the *meaning and character* of the entire event,[71] to the extent that – although it is perhaps possible to name a *common line* – the prevalence of either biopolitics or democracy appears as decisive. For this reason, in *The Rebirth of History: Times of Riots and Uprisings,* published after his article on Libya,[72] Badiou practically reflects exclusively upon events

---

69 J.L. Nancy, "What The Arab Peoples Signify To Us", *Liberation*, March 28 2011.

70 A. Badiou, "Alain Badiou's Reply To Jean-Luc Nancy", www.criticallegal-thinking.com Also see T. Ali, "Who Will Reshape The Arab World: Its People Or The US?", *The Guardian*, 29.04.2011.

71 In the collection of essays *The Dawn Of The Arab Uprisings* many authors refer to sociological and political differences between events in Arab countries. See B. Haddad, R. Bsheer, Z. Abu-Rish, R. Owen (ed.), *The Dawn Of The Arab Uprisings* Pluto Press, London, 2012.

72 Certainly, Badiou cannot be accused of any inconsistency here: in conclusion of *The Rebirth Of History*, he clearly states that on Arab Spring he has previously published two articles i.e. one on Tunisia (*Le Monde*) and the second one on

in Tunisia and Egypt, as the examples which are certainly the closest to authentic people's riots. However, claiming that the *figure of riots* is still not a *political figure*, and that only when the time of the interval is over we can speak of new politics, Badiou explains that we already know – especially in Tunisia and Egypt – the protests will continue, but also that they will *split*.[73] This statement affirms our previously articulated thesis: the mentioned riots oscillate between the external impact, *ergo*, neoliberal interests and people's uprising, *but can still not be called historical uprisings* because of their lack of a *binding idea* which causes *splitting*. Alternatively, to the extent in which a riot is capable to articulate and bring into life an *authentic ideology and new politics* it will correspond to the name of a *historical uprising*.

No less relevant, the issue then also becomes how valid is Badiou's historical parallel between the Arab Spring and 1848 Europe, i.e. are we dealing with "the same general uprising...the same anti-despotic orientation, same uncertainties".[74] However, one should have no dilemma in respect to the special character of Badiou's theory, which certainly cannot be regarded as a part of the produced *mainstream* Western discourse on "Arab revolutions", i.e. of a discourse that in a hegemonic manner simplifies these events, reducing them to despotic regimes vs. fights for freedom. Because, in the work *The Rebirth of History* it is emphasized that the so-called *"friendly countries"* – such as Saudi Arabia, Pakistan, Bahrain, Nigeria, Mexico and many others – are certainly no less despotic or less corrupted then Ben Ali's Tunisia or Mubarak's Egypt have been, and often they are even more corrupted. This is, let us say, equally one of most relevant differences between Badiou's and Žižek's analysis, because Žižek, completely in harmony with the *mainstream interpretations*, explicitly

---

Egypt (*Liberation*). Badiou explains that his attitude regarding the two events has been completely different: while in the first case he attempted to articulate the moment of universality in the uprisings, in the second he took a highly critical stance towards the French-British intervention. See A. Badiou, *The Rebirth Of History*, p. 103.

73 *Ibid.*, p. 47.

74 *Ibid.*, p. 48. A similar comparison with 1848. has been made by Eric Hobsbawm. See E. Hobsbawm, "It reminds me of 1848...", by Andrew Whitehead, BBC 23.12.2011.

speaks about the *"revolutionary spirit"* of the Arabs, emphasizing that this is in line with "the best secular democratic tradition", i.e. that people protest against "repressive regimes", against "corruption and poverty", and that he hopes such protests will spread to countries such as Saudi Arabia.[75]

In such a way, a discourse which, insisting on spontaneity and originality, attempts to leave behind the entire dimension of Western interventionism, and consequently Western interpretations of the so-called Arab Spring, is constructed. Moreover, in difference to Žižek, Badiou is clear in stating that only if the desired explanation of the West – that Arab Spring is the *desire for the West* – turns out as completely untrue, i.e. if the inclusion in this sense does not succeed, can we speak about authentic change, because it means an exit from the West, a de-Westernizaton and exception.

In this light, it is indisputably necessary to turn to Hardt's and Negri's characterization of the Arab Spring, which resembles a great deal to Žižek's articulation. Moreover, these authors attempt to say that the multitude is politically subjectivated not only in riots in Tunisia and Cairo, but in Libya as well. In saying that "the multitude is capable of organizing itself without a centre" and that "the insurrections of Arab youth are certainly not aimed…at a form of democracy adequate to the new forms of expression and needs of the multitude."[76] Hardt and Negri, as Nancy or Žižek – unlike Badiou – insist that, always, and practically unanimously, we are dealing with completely authentic events which refer to a new birth of democracy.

The question remains with how much argumentation the Arab Spring appears as an *event,* in Badiou's sense of this concept,[77] i.e. to what extent can one actually speak about a historical uprising *par excellence.* In detailed analysis of events in Egypt, Badiou refers to statements and inscriptions from the Tahrir Square, that were mostly concerned with

75 S. Žižek, "Why Fear The Arab Revolutionary Spirit?", *The Guardian*, 1.02.2011.
76 M. Hardt, and A. Negri, "Arabs Are Democracy's New Pioneers", *The Guardian*, 24.02.2011.
77 More on the philosophical concept of event see A. Badiou, *Being And Event*, Continuum, New York, 2006.

"the state", "Egypt", "return of the state to the people" and "the right of Egyptian people".[78] In other words, these messages were conceived to point to the end of servility to the West, as well as to the end of radical inequalities in society, but, on the other hand, this is still highly insufficient for saying that we are faced with a true leftist movement here. However, doubtlessly the articulation that the *change of the world* is possible only when people who are *absent* from it *appear*, i.e. *the inexistent*, reminds us that exclusive significance lies precisely in *the rise of the nonexistent,* since only this can we call *a historical riot*. The same goes for perceiving that in these riots, such as in Egypt and Tunisia, we see something of these ruptures and of new invention of time: an opening in which all nonexistent can potentially appear, *ergo*, an opening of political truth that emerges from the authentic event.

Here we should turn our attention to two issues emerging from Badiou's analysis: the first is that, for Egypt, but elsewhere too, a lot will depend on the outcome of the fight for *the new constitution*; the second is that this condition, although necessary, is still insufficient for change to appear in reality as real transformation. Finally, Badiou is right to insist that the key moment lies in the following: to what extent is this going to be a self-determination of the people since the people in Africa, as well as Asia and South America have carried on their backs for a long time the hard burden of Western *colonialism*. This way, Badiou emphasizes that, firstly, the dilemma is one between biopolitics and its end, and, secondly, that majority of these people do not want war, nor are scared of it, but are, however, ready to risk their lives for radical changes.

Simultaneously, this is maybe the first point in which different meanings of the Arab Spring, torn between biopolitics and democratic uprising of the people, appear. Because, to the extent in which there is readiness and decisiveness in the struggle for justice and freedom, i.e. for the ideas which are priceless, we can speak about a leap from neoliberalism and about the birth of a new epoch – potentially about new forms of rebirth of history as well.

---

78 A. Badiou, "Tunisie, Egypte: quand un vent d'Est balaie l'arrogance de l'Occident", *Le Monde*, 18.02.2011.

The Occupy Movement – Simulation Or The Initial Step Of Resistance?

If we have attempted, in our previous chapter, to sketch the philosophical issue of the so-called Arab Spring as a contemporary phenomenon with numerous potential implications for the rethinking and practice of *the political*, this questioning, on the other hand, has already led us to the significance of articulating another manifestation – the Occupy Movement as a specific form of *Western riot* against the existing order. Although of European origins, the appearance of such a *movement* has its most notable shape in US gathering, i.e. in the collection which took place *in the heart of the Empire,* and therefore, our attention will mainly be directed to this *location.*[79] For if, on the one hand, the Occupy Movement clearly had its exemplary inspiration in the Arab Spring[80] – and in such light is conceptualized to represent a form of *the rise of the West,* i.e. a *Western response to* the *biopolitics of the West* – it is simultaneously clear that its most famous form is precisely the *Occupy Wall Street* (OWS), as the event in the heart of contemporary capitalistic politics.

Doubtlessly, reactions and interpretations of this riot vary from uncritical glorification and the statements which referred to it as the crossroads of world events, to the refutation of OWS as a scam and simulation organized by the system. Such reactions appear in one aspect as *passion* with the respect to the *desires for authentic transformations,* much more than as conclusions of objective analysis of causes and effects of this movement.

---

79 This certainly does not mean that we will neglect the *de-territorializing aspect* of this movement, but that there is something of a particular *relevance* about Occupy in the *US*. Moreover, for contemporary Europe, we are faced with specific phenomena which we can call *critical European discourses,* and which cannot be reduced to this movement. This will be subject of analysis in our next chapter.

80 In OWS programmatic statement it is written that this "is a leaderless resistance movement with people of many colors, genders and political persuasions. The one thing we all have in common is that We Are The 99% that will no longer tolerate the greed and corruption of the 1%. We are using the revolutionary Arab Spring tactic to achieve our ends and encourage the use of non-violence to maximize the safety of all participants." www.occupywallst.org.

Or, it seems to us that, here, perhaps, we should first direct our attention to the *discursive and practical implications,* as well as to the moment of the given context, i.e. the *specific localization* in the heart of the US. In this light, the stress on the actual occurrence of something that was *unimaginable* only a few years earlier, even as a *possibility,* in *place* and *time* of neoliberal hegemony, opens – in a subtle sense – the horizon for understanding that, provisionally speaking – with all more or less evident insufficiencies and lacks – we are dealing with a *specific event,* with a *trace* of which gradually opens the *potentiality* for *the political.*

Here we should notice a significant difference between contemporary phenomena of the so-called Arab Spring and the Occupy Movement, which is based on the difference of location and, consequently, on the influence of localization on the defining the potentials for the political. In such a way, simultaneously with the articulated theory of time of the political, a theory of the place of the political appears as well. Or, if in the Arab Spring "the riot against" still turned out to be insufficient, i.e. if, here, what was needed was a clear *"yes!"* so that we could speak about authentic politics and historical transformation, for OWS – since we are speaking about a protest within the heart of the Empire, the sense and significance of "the rising *against"* appears as a specific theoretical, practical and normative value *as such.* Conditionally speaking, therefore, if the *Occupy Movement* had not succeeded in anything else, it has already marked a relevant turning point by the fact that *class* in US has appeared *as a political name.* Through a symbolic parole *"We are the 99%!"* the movement referred not only to economic inequalities but also to the *structure of the whole neoliberal system,*[81] whereas, on the other hand, it spoke against *political representation* as well.[82] So consequently, the OWS, in spite the fact that it

---

81 In this sense, addressing OWS, Naomi Klein emphasized that the movement chose *"the right target",* while *"courage and moral compass"* represent key virtues of such a riot. See N. Klein, "OWS: The Most Important Thing In The World Now", *The Nation,* October 6, 2011.

82 Precisely OWS developed the organizational structure that opposes representative democracy: the central role of "the general assembly" where decisions are made and which is open for all. However, as Cinzia Arruzza pointed out, the fact that decisions are made on the *principle of consensus* testifies that what is still lacking is the articulation of *immediate democracy,* since *democracy here*

did not have a defined program and was characterized by a multiplicity of ideas, has demonstrated two significant things: that the *social issue* in the US is finally recognized in a serious way; and that it has articulated that the *concept of democracy* has something essential to do with *the people* and the *government of the people*, i.e. with *polis* which is *something else* in comparison to existing *mimesis* and *virtuality*.[83]

This is precisely the aspect revealing that such a movement cannot legitimately and objectively be regarded as mere simulation – unless it is already presupposed that every critique of the system is in the service of the system. For this, consequently, would make *any critique always already irrelevant, worthless and suspect at the same time*.[84] Moreover, it seems that the theoreticians and activists who emphasized that any insistence on strict and clear ideological frameworks would only limit the movement – since this is exactly what the critiques of the OWS most insisted on – are right in one significant aspect.

But let us return once more to the relevance of localization, i.e. to the location of the movement in the heart of neoliberal, *ergo*, biopolitical hegemony in the US. Apart from the fact that, as we have already noted, this type of recognizability and placing refers to one condition for the appearance of the political (for there is no *political without location*), it simultaneously makes clear why all objections claiming that the Occupy Movement actually affirms colonialism or neo-colonialism are practically completely unjustified. Because, the critique claiming that participants of this movement and participants in the protest did not desire the position

---

is replaced with consensus. C. Arruzza, *Journal For Occupied Studies*, www. occupiedstudies.org/.

83 The fact that US academic institutions, such as Harvard and The New School, as well as many respected intellectuals, organizations and individuals, have supported OWS – and that, no less significantly, that the movement also began to spread *horizontally* – indicates that both political and social implications of this riot are yet to be manifested in time to come.

84 Naomi Wolf has published an article in which she reveals the documents of American government that disclose that FBI and DHS have followed the OWS as a common terrorist force in spite the fact that it was officially proclaimed to be a peaceful movement. See N. Wolf, "Revealed: How The FBI Coordinated The Crackdown On Occupy", *The Guardian*, December 29 2012.

of the US as a colonial centre to be threatened,[85] overlooks precisely that here we are dealing with a specific – and *specifically Western* – discourse. It is an attempt to remind us of the existence of values such as justice and government of the people, and it is precisely trying to do so from the place where destruction of these values began. And this fact, to say the least, is not insignificant.

Articulating the purpose of the movement, Noam Chomsky has written that the contemporary *division of the world*, in terms of the Occupy Movement, is the division between the 1% and the 99%. Chomsky emphasized that "this movement is the first big reaction of the people which can overturn this", but also that "victories are not made tomorrow".[86] In this respect, the Occupy Movement raised the question of social and economic inequality, i.e. it began questioning the issues of the distribution of financial means in the US, pointing to radical inequalities For Chomsky this raises important the questions about the system, and, simultaneously, refers only to the beginning of a struggle which will, in different forms, last a long time. Also, keeping in mind Badiou's criteria according to which one condition for real riot is that its participants come from different social and other backgrounds, the perception that the carriers of the Occupy Movement are mostly members of the *middle class* becomes even more relevant And what is even more important is the role of *the young, the intellectuals* and even of some members of higher class in it.

The simultaneously localized, and socially contextualized subjectivity testify to the growing consciousness of radical inequalities, among the population whose existence was not in question. Although inspired for peaceful protests, at least some of the participants were ready even to fights in the name of the *unnamed*. Certainly, it is completely clear that in OWS we have not seen material, i.e. real mass movement, one which is necessary

---

85 The point here would be to say that the protesters are not led by the attitude that it is solely them who should begin a world revolution, i.e., that there is no exclusivity in comparison to the non-Western world, on the basis of which it would be plausible to say that they adopt a neo-colonial stance.

86 N. Chomsky, "If We Want A Chance At A Decent Future, The Movement Here And Around The World Must Grow", Howard Zinn Memorial Lecture, 3 November 2011. www.alternet.org.

for a serious reversal. On the other hand, however – and regardless of how paradoxical this might sound – the aim has been, first, to show the existence of consciousness and awareness about this social gap, i.e. to simply refer to the moment of becoming of the political, which consequently also explains the slogan that the protesters *do not know what they want.*

Taken together, these two moments – of which the first draws out the fulfillment of one of *causa sine qua non* of Badiou's theory of authentic riot, while the other reveals the nature of something similar to an *emancipatory political experiment* – disclose the theoretical-practical character of a contemporary phenomenon which, apart from everything else, is directed towards strengthening the role of public gatherings, as a form which affirms the existence of the inexistent, i.e. the voice of the people. On the other hand, indisputably, this is not enough to announce the *revolutionary* character of the movement, in the same way in which for the so-called Arab Spring it is still uncertain whether it will appear as a historical uprising *per se.* Now, it becomes clearer why the differences between the two concern the specificity of their location but equally of their aims, and because of this, the *conditions* that are placed forward regarding these two phenomena of protest *are not the same.* But the *potentiality for the political,* present in both contemporary events, appears precisely as bringing into life the *possibility of riot,* which, consequently, transforms the social-political setting and, in demonstrating *the power of the voice of the people,* represents the first step toward *immediate democracy.*

In this sense, Critchley's observation that "the Occupy Movement and the Arab Spring will continue to *revitalize political protests,*"[87] appears precisely as the moment that refers to a particular process of strengthening of the political impulse, i.e. of the intensification of the authentic appeal for participation and innovation in a different theoretical-normative framework. On the other hand, although this insight *eo ipso* still does not mean the affirmation of the attitude that, for example, the OWS movement has made a substantial leap towards "the renewal of direct

---

87 S. Critchley, "Occupy And The Arab Spring Will Continue To Revitalize Political Protest", *The Guardian,* 22 March 2012.

democracy",[88] it refers to the moment which, in Badiou's terminology, is properly philosophical, but which, as such, points back to the relevance of political-philosophical discourse for political practice.

Reminding us of Bauman's articulation that in the contemporary world *power* and *politics* are separated to the extent that power today stands on the side of technocratic crypto-elites, Critchley explains the sense in which *"politics today has no power, but serves power"*.[89] This is the reason why the return of the political first appears as the necessity of confrontation with the power of financial capital and, therefore, as the space in which the relevance of the Occupy Movement comes forth. As regards the separation between politics and power, this movement symbolically outlined the potentiality of joining on an entirely different basis, which attempts to refer to *the autonomy of the political*. So, to say that the Occupy Movement signifies a becoming-conscious of deep dissatisfaction with normal politics, especially among the young. As such it is a phenomenon of "politicized, radicalized youth – after almost two decades of postmodern irony"[90] and it means to bring into light its political-philosophical truth implications.

If *politics* conceptually originates in concepts such as *justice* and *freedom*, consequently, each time a new horizon of this interrelation opens – even as a yet-to-be-articulated possibility – a *contemporary leap from biopolitics* takes place. In this light, if it is kept in mind that the lines of the thought that *capitalism* is a political and social name which *needs to be abandoned*, this may be done in the name of something still unclear and new – and in this sense the relevance of the Occupy Movement certainly surpasses its immediate, and immediately visible, implications. Or, although it is indisputably true that one can speak of "world revolution" here, a revolution which would, in an ideal-type manner bring forth the dissolution of class differences and the end of exploitation, we cannot completely cast aside this movement by classifying it as merely an antiglobalistic carnival or art show. For this would mean not to recognize its

---

88 M. Hardt, M. and A. Negri, "The Fight For Real Democracy At The Heart Of Occupy Wall Street", *Foreign Affairs*, September-October 2011.
89 *Ibid.*, p. 76.
90 *Ibid.*, p. 79.

theoretical and practical scopes, which refer to the attempt of a new creation of the political, i.e. to an entirely different aim compared to the issue of organized political struggle.

In this sense, Ranciere articulates the Occupy Movement responds to *"the fundamental idea of politics"*. *Power* is given "to those that are by no particular motive directed towards *governmentality* ... and they have *materialized this power* in such a way that corresponds to this fundamental idea: affirming *the power of the people* through *subversion of normal distribution of places*: normally there are places, as streets, intended for circulation of individuals and things, and public places; such as assemblies or ministries, intended for public life and dealing with common affairs. *Politics is always manifested though perversion of this logic."*[91] Or, more precisely, if the stress is placed on the moment that contemporary political parties, as well as most of the practices of the so-called "normal politics", are reduced to the usurpation of power, the birth of the political, consequently, appears in the light of the events that produce ruptures in this logic.

This also implies the creation of different places for public debates; the flow of information and establishment of new forms that contribute to the growth of autonomous thinking and action. Moreover, Ranciere reminds us that "the point is not in how much philosophers will be present in these gatherings. The point is the existence or nonexistence of world image that naturally structures common action. In May 1968, although the form of the movement was far from the canon of Marxist politics, Marxist explanation of the world has functioned as the horizon of the movement. Although they were not Marxist, "they have situated their action in vision of history in which the capitalistic system will disappear...and *today's demonstrators no longer have the horizon that provides the historical validity to their struggle...*"[92]

But, precisely in this lies the relevance of the Occupy Movement which, in spite of numerous contradictions, essentially reminds that *power of the citizens* is precisely their power of autonomous judgment and action. Or, it

---

91 J. Ranciere, "Hablar de crisis de la sociedad es cuplar a sus victimas", *Publico*, 15.01.2012.
92 *Ibid.*, p. 2.

reminds that every form that affirms the existence of the capacity for *anyone to judge and decide,* i.e. every form of public activity, joining and action, which is first directed against monopolization of technocratic elites, brings into light the sense of the *first democratic virtue per se,* the virtue of belief in the capabilities of anyone, as an equal participant in *polis.*

The Birth Of Radical European Politics Of Left And Right

A particular phenomenon, which should be considered in the discussion of contemporary biopolitics and the attempts to, contrary to neoliberal hegemony, reaffirm *the political,* is the emergence of divergent processes which are growing in different European countries. This is to say that simultaneously with manifestations which have exemplified that the crisis of the European union is not simply *economic* but that we are facing a systemic crisis which, in a significant sense, is a *political and social crisis,* it has become clear that the same, seemingly contradictory impulse, of simultaneous strengthening of mechanisms of control and domination, and gathering which confronts it, is also taking place on European grounds. Moreover, Europe is becoming the exemplary area of such developments. Because, to the extent in which in previous times European citizens were faced with the growth of consciousness that the EU practically has nothing in common with its own proclaimed slogan about *equality of the people* and *equality in difference,* they are becoming aware now of the gap between the *EU crypto-elites* and the values such as *democracy* and *freedom.* The gap – which we can call the gap between *EU forms* and *European realities* – is growing in its multiplication and expansion. And so we have found ourselves *in medias res of contemporary biopolitics.*

First – as Ranciere already noted as early as in 2005, explaining why the French will say "No!" at the referendum on European constitution – the EU appeared as an exemplary case of simulation of politics, i.e. as an example of political virtualities which are first recognized through the establishment of the so-called consensus which, by its structure and its approach to the essential issues, *erases the difference between the conservative governments and socialist oppositions or vice versa.*

Referring to the French refusal at the referendum, Ranciere wrote: "How is such a thing possible, it was asked, when both the *conservative government* and the *socialist opposition* called to vote 'yes'? This is because, came the response, the French have not understood. They want to express their discontentment with their government... What merits our attention is what is expressed by the reasoning of its doctors – this medicalization of opinion, this interpretation of every vote that does not conform to the official expectations as an expression of a pathological state. If an electoral body is asked the question of whether it is for or against a measure proposed by its government, then the proposition must actually include the possibility of a negative response...So why is there so much surprise and desolation when the free, unpredictable choice included in the rights granted to citizens is actually translated in act or threatens to be as an unforeseen response?"[93]

This way, Ranciere's political-philosophical reflection places us *in medias res*, in the heart of the *European problem*, opening the path for what we can call today *contemporary critical European discourses*. From 2005 to 2014 it has become transparent that many governments of the so-called "democratic regimes" of European states are more loyal to the idea of the *EU*, than they are to the idea of respecting *the voice of their own citizens, the voice of the people* (and, for that matter, the European Constitution, to which the French said "no", still exists). It is a situation in which you can talk, and scream, *but no one will hear you*, the voice of the people is like *a cry in a desert*, while the sand is getting deeper. This is why *democracy of the EU* is *an event which has not happened*, an *invented event* which *has never been realized, a non-event of non-location*. Moreover, it seems that the choice of defending *EU ideology* – Marx spoke of *The German Ideology*, while today we can speak about *European ideology* – has been made even at the price of *the disappearance of the concept of politics*. Such *anti-political* and *anti-theoretical* impulse lies *at the core of the project of the EU*. European ideology is the opponent of Europe of knowledge since, in order for domination to be complete, it attempts to rule out the possibility of resistance that rises from dialectics of power.

This is the key reason why, in the course of time, both in theory and in practice, different articulations of *contemporary critical European discourses* appeared. Although they moved in completely different directions of *left* and *right*, these discourses manifested themselves as *radical politics*. In both cases, politics confronts quasi-democracy, stressing the need to hear the true voice of the people of European countries. In such a way,

---

93 J. Ranciere, "Democracy And Its Doctors", *Chronicles Of Consensual Times*, Continuum, New York, 2010. p. 141–142.

explicitly or implicitly, this showed us the EU turned out to be an exemplary anti-political and anti-philosophical project. Moreover, the thesis which we will attempt to articulate is the following: the systemic production of depolitization in Europe in last decades had for its aim the foundation of biopolitical, *ergo* neoliberal, control of the entire populations by *trans-European crypto-elites*; and *sine qua non* of such a project has been the strategic positioning of anti-European discourse as inherent structure of the existing Union, therefore, of a discourse which would intentionally remain *beyond liberté, égalité, fraternité* – and at the same time it would gradually distance itself more and more from the Aristotelian *koinonia politike*.

In this light, the diagnoses such as Habermas's or Sen's,[94] which emphasize that the key problem of the EU is the so-called "democratic deficit", i.e. the fact that the Union has entered the "*post-democratic era*" and, consequently, all analyses which do not bring into question the system itself, and in principle can be reduced to the appeal for its improvement, appear as insufficient. Therefore, it seems that Ranciere's analysis that has focused on the articulation of the *lie of representativity*, i.e. on the representative model of democracy in contemporary Europe is far more instructive here, since it has shown that we are dealing with authentic insufficiency which

---

94 See J. Habermas, *Zur Verfassung Europas*, Suhrkamp, Frankfurt, 2011. In spite of the fact of monopolization of the European project by self-proclaimed elites – as the cause of bureaucratization – Habermas is still faithful to the project of the EU, with the emphasis that giving up on European unification would mean partition from participation in world history. In other words, although he claims that the EU has transitioned into a *post-democratic era*, Habermas is focused on democracy exclusively from the perspective of "the democratic deficit". From this perspective, it is possible to speak about lacks in matters of procedure, such, for instance, that EU representatives, except for European Parliament, are *appointed* and not *elected*, and that in the case of *EU representative democracy* it is not clear *who do EU representatives actually represent*. Like Habermas, Amartya Sen articulates that "perhaps the most problematic aspect of current European diseases is the exchange of democratic obligations with financial dictates – by EU leadership and European central bank", and that "Europe cannot surrender itself to unilateral understandings – or good wills – of experts, without public debates and informed consent of its citizens". See A. Sen, "The Crisis Of European Democracy", *The New York Times*, May 22, 2012.

reflects *the disappearance of the political*. According to Ranciere, the lie of representativity happens in a twofold sense: representativity in particular states, as well as on the level of the Union, because in both cases *the people* who elect the representatives are in principle *missing*. One of the causes of such a situation is that what is opposed to the idea of *free autonomous choice* is the so-called *knowledge of the experts* which has been practically mythologized, presented as an unquestionable value, worth more than any and all particular voices.

This is not the old Platonic dilemma about philosophers-kings, it is not the question of whether the most *knowledgeable* should rule or not – because *techno-managers of the EU* are not carriers of such knowledge. Furthermore, it is precisely these *experts* who have emerged as the greatest opponents of *knowledge,* placing themselves in the position of *crypto-pedagogue* and *therapist*, through a series of contemporary techniques, attempts to bring a primitive patient to the stage of consciousness.

Ranciere writes that in most radical cases this exercise is transformed into *"a cruel psychoanalysis of the ill social body*. This is the relevance of simulation of polls and the manifestation of enormous interpretative work that governments, experts and journalists undertake to demonstrate to *the people* that they are *a sick population* if they believe that they can *choose*. In this sense, Ranciere emphasizes that *the European referendum has brought this logic to daylight*. If we were to describe in more detail the tragic and comic character of this logic, then we would see that the choice between the *EU*, on the one hand, and *Europe-which-is-not-the-EU*, has been presented by *Eurotocracy*, the *class of Brussels*, which is part of the model of *universal monarchy* and rests on acceptance of such an order – practically as a choice between *being* and *nothing*, between pure life and violent death. This is most obvious in all statements articulated in the following framework: if you do not say "yes" to the EU, then you do not have an "affirmative" but a "negative" attitude that in final instance leads to self-destruction. And what *does not exist in such discourse* is precisely *Europe*: instead there is *Eurocracy* and the *oppressing of the people*.

The problem of the EU, and of its contemporary and growing *political crisis*, therefore, does not consist merely in the *lack of democracy*, but also in the original insufficiency which refers to *the disappearance of the political* – because the concept of politics always presupposes *the moment of subjectivity* and *the moment of democracy.* This certainly means that the project of the EU is not in crisis because the monetary union has structurally not been conceived in a proper way (although it is most likely this is the case as well).[95] Consequently, the essence of the problem lies in what Badiou has termed as *the materialistic paradigm*[96]: in discursive and political forms of *neoliberalism and postmodernism* and their production of *post-politics.*[97]

However, returning to this lack of *political subjectivity* – it is not only a matter of the fact that the EU has not founded common foreign politics (diplomacy, security, defense), and has just partially succeed in internal matters, but, first of all, that is has not established a bond between people, such that would create a sense of togetherness, which is *why Europe, in various political and cultural aspects remained beyond the EU.* To what extent has the European Union been imagined as to be a live example, a *par excellence* example of *postmodernity* is perhaps best expressed in explicit statements and writings of one of the founders of the *EU ideology,* practically as a new religion, Robert Cooper. Cooper claims that "the EU is the most picturesque example of the postmodern system". [98]

If we want to take this thought further we could say that the EU is a *substitute,* a *replacement for the subject,* that the stage of EU ideology is such that the main play on the scene is one where *the subject is missing,* that *the subject is absent;* the key role belongs to the *EU* – but the *EU is not a subject.* So it is a situation of paradox, where the subject which formally is a carrier of legislative, executive power, of economic and social power – *is actually not a subject* (it is situated in a space between a *replacement* and *supplement of power, a subject-fiction,* a *virtual subject* operating in the mode of "*as if* ".). On the other hand, the time of "triumph of

---

95 See T.G. Ash, "The Crisis Of Europe", *Foreign Affairs*, September/October 2012.

96 A. Badiou, *Polemics*, Verso, New York, 2006.

97 Speaking about the left i.e. about its 1968. "sins" in this context, it would be possible to consider the idea about "the infinite enjoyment", of "our wishes" as "the reality" – as the signification of transition from *bourgeoisie capitalism* to *post-bourgeoisie capitalism.*

98 R. Cooper, *The Breaking Of The Nations*, Atlantic Press, New York, 2003. p. 150.

the Union" symbolically and chronologically has been synchronized with the era of triumph of neoliberlism and contemporary capitalism, which was supposed to bring the so-called "end of history", and with it "the better life" through free flow of people, commodities and capital.

In this respect, there are structural similarities between *EU bureaucracy, Eutocracy* as a specific *class* that, has been spreading, for decades, sometimes hysteric, *Europhilia, as an ideological fiction,* and Marx's description of the *bourgeoisie* in *The Communist Manifesto.* Similarity appears precisely in their common *liberal aspect,* i.e. in the fact that, like EU bureaucracy, "the bourgeoisie has left remaining no other bond between man and man then *naked self-interest* and callous "cash payment"... It has resolved personal worth into *exchange value,* and in place of the numberless indefeasible chartered freedoms, has set up that single – *free trade."*[99] *Eurotocracy* has brought with itself a theoretical, normative and existential fall, deepening the gap between the rich and the poor in a most radical way.

Now, the question is what are the causes of these processes, that have been, in their contemporary outlook, followed by relativization of key concepts of politics, such as, *the people, justice, equality*, while concepts such as *democracy, freedom, citizens, elections,* have been preserved but *emptied of content* (this has further enabled the arbitrary use of such concepts); while concepts such as "peace" ("peaceful unification of citizens"), "standards", "development", "better life", "prosperity" have been used to the extent that they also appeared as empty signifiers? How did this happen? The main cause is the *consensus of crypto-elites,* that effectively erased the differences between the left and right, induced an atmosphere of *sterility,* of a space *emptied of content* and practically, this way ruled out the possibility of originally diverse political choice.

In the situation in which everything has appeared as *the other of the same*, in two or three shades of grey, it has to be noted that the institutions of the European Union were built in a way that opposes European democratic traditions (a typical example of this is the European Parliament, that

---

99 K. Marx, *The Communist Manifesto*, Pocket Books, New York, 1964. pp. 61–62.

had very little power for years, while the case of other EU institutions is even more obvious). In other words, the bottom line is that the causes of the contemporary situation in Europe are *fake politics*, a *fake consensus* in which absolute capitalism cannot be questioned – the *political simulation* in which both left and right lost their own political identity.

Therefore, when he wrote that *the EU is the most developed example of the postmodern system*, Robert Cooper actually emphasized precisely what has turned out to be the key characteristic of the EU project, and that is the relativization of concepts of *politics* such as *the state, people, law, sovereignty*, as well as the marginalization of categories of *justice* and *equality*. Simultaneously, the articulation and practices of contemporary Union attempted to preserve concepts such as *democracy, freedom, citizens, elections*, but in such a way that they become an *empty place, ergo, emptied from any content*, with the aim of making them available for *arbitrary (mis)use*. In a situation in which the consensus of crypto-elites has already been made everything appears as the other of the same, since the potential for authentically different political choice has been abolished. This type of *applied postmodernism*, in relation to *European neoliberalism*, has secured an almost total domination of *the materialistic paradigm*.

In such a theoretical-practical cross-cut a particular *class* has been formed, the class made up of EU bureaucracy and its employees, which spreads *the irrational hysterical Europhilia*. Their *Weltanschauung* brought to light its high proximity to the description of the bourgeoisie from Marx's *Manifesto* for it, equally, has not any other connection between man and man aside from *naked self-interest* and has turned personal value to *use value*, and on the place of multiplicity of freedoms left only one – *free trade*. The transition from the EEC to the EU, as the transition from *economic* to *political* community, has been exemplified in Monnet's persuasion that *technocratic measures of economic integrations* are sufficient for the political. This has, therefore, followed the conception of the EU from its beginnings, for the Union was supposed to symbolize an existential crossroad, that expands to all spheres of politics, society and personal life. The idea of the disappearance of the political has been, therefore, basically incoherent from the start, because it did not foresee that without *political identity* or, rather, *political subjectivity*, the EU cannot be sustainable, in the similar sense in which *simulation of democracy* cannot erase the fact that *the people exist* – and that in the *dialectics of de-politization* what occurs is precisely *re-politization in opposite directions*.

In this respect, the fact that the EU has not manifested *"unity in diversity"*, i.e. that it has not achieved even a basic agreement about its *political identity*, nor has it turned into reality the slogan of *"equality of people, freedom and prevalence of reason"*[100] – is further radicalized by its *crypto-ontological inscription* in which, *ergo*, it has turned out that what is at stake is the interrelation between *liberal* and *postmodern project*. So the birth of contemporary critical European discourses and, simultaneously, the strengthening of radical European politics of left and right has appeared as a natural and logical consequence of dogmatization, irrationality and mythologization.

In time, the EU has grown to be a *counter-example,* an anti-model, all of which makes Balibar's argument that "Europe is a dead political project" because "some countries are *dominant*, while other are *dominated*" more and more convincing.[101] Or, more precisely, the gap between the EU *crypto-political elites,* as a specific superior *class,* and *people,* has grown to the extent that it has caused differentiation on both the *social* and the *national* basis, i.e. it has marked the relevance and urgency of the issue of *equality* and the issue of *identity*. Because, as consequences of the process of the establishment of a virtual political community, i.e. of *political community without political subjectivity,* which was simultaneously supposed to mark the destruction of both *left* and *right*, completely new and different theoretical and practical discourses came into being. According to the compelling articulation of Tariq Ali, the reason for *EU neoliberalism* is the fact that Europe has become *"the prisoner of Atlantism"*, which has caused the weakening of critical thinking and political imagination.[102] This is why the common signifier of new politics appears precisely as

---

100 Thomas Meyer, *Die Identität Europas*, Frankfurt am Main, Suhrkamp Verlag, 2004.
101 E. Balibar, "Europe Is A Dead Political Project", *The Guardian*, 6 June, 2010.
102 T. Ali, "Evropa je zatočenik atlantizma" ("Europe Is A Prisoner Of Atlantism"), *Večernje novosti*, 6 Septembar, 2011.

Euro-realism and/or *Euro- skepticism* which represents the appeal for systemic transformation of discourses and practices, i.e. the appeals for *political, theoretical, social and economic changes in European states.*

John Gray has written that if its architects had imagined the EU as a new model for the world, the current situation points to an opposite twist. This is to say that precisely *dogmatization* and *mythologization* of the EU has conditioned its appearance as a *counter-example,* an *anti-model* and, further, that *new fragmentation*, primarily on *social*, but also on *national* level, brings the contemporary condition close to Balibar's radical insight. This means the following:

(1) first, the gap between the *EU crypto-political elites* and *the people* has dramatically grown in the last few years;

(2) second, this gap has signified a deepening of *social* and *economic in-equalities*, which is why *class* again reappears as a relevant political name; (in the EU today the level of poverty has reached 120 millions of population, and is estimated soon to reach an even more dramatic stage, of almost 1/3 of the population of the European Union);

(3) third, because the national issue has been treated by EU elites always as a question of *Fascism* or, in best case scenario, as *passe* and completely irrelevant, the result is that nationalism is returning through the back door.

In other words, the fact that *radical politics of left and right* are being re-born and self-produced, has its basis in the paradigm of the EU, in the choice of its concepts, and its practices, which exemplified an end of politics of both left and right, aiming to create a *contradictio in adjecto,* i.e. a *political community without political subjectivity.* How the destruction of the left in a framework of Blair-Schroeder relation, i.e. as so-called politics of *the third way* (presented as "*social-democracy*") has been taking place, though the formula which was supposed to say "*beyond left and right*" – has been articulated by Chantal Mouffe, while Robert Cooper's *Breaking of Nations* is a irreplaceable example of plea for post-national post-modern political order of the EU. The point is, therefore, that in over two decades the EU has been built as a *political simulacrum* – because its so-called "left" and "right" politics were realized *as simulation of those politics.*

So, the first decades of the 21ˢᵗ century – with spreading economic, political and social crisis – in theoretical, but simultaneously in a very practical sense, have brought the rise of *Euro-realism* and *Euro-skepticism* precisely because the complexity of the *EU crisis as crisis of the political*, at the same time appeared as *the crisis of political participation, the crisis of democracy and the crisis of political subjectivity,* that is, as *the crisis of legitimation of the EU project per se.* This is why the idea of necessary transformations is growing practically everywhere in Europe – because of *politics, philosophy, law* and *history*, but also because today even bare life i.e. the sphere of *bios* has been brought into question. It is relevant to keep in mind here that EU ideology has, for the most part so far, either *criminalized* its opponents, or *corruptionalized* them, or simply reduced them to "forces of the dark". This way, EU ideology has departed from the dialectic between the particular and the universal, forgetting that any attempt to negate either of these elements ends up in negating the other one as well. And, no less relevantly, there is a point where *struggle for bread* and *struggle for freedom* meet: the struggle for *survival* and for *dignified life* appear as inseparable parts of the same process.

This fact that the whole sphere of *bios* is at stake now also means that *contemporary power of governing* has departed from the sphere of its *invisibility*, surpassing even the situation which Foucault described as fragmentation of power – the situation where power becomes capillary, spreading to institutions and different forms of life. This means that the subjects of imperial and colonial power of neoliberal governmentality are no longer concealed. Therefore, biopolitical governing as governing of entire populations can, for example, be traced in the *normalization* of repressive social practices, in the articulation of scientific and cultural discourses in which freedom is reduced to arbitrary freedom of choice in the framework of "the same", while concepts of justice and equality appear as relicts of the past. However, the fact that power of *Eurotocrasy* – as a *class* whose interests lie beyond the interests of *the people* – has now turned to *absolute visibility* means that *the Europe of the poor* has become transparent. Moreover, it is becoming indisputable that such *Europe of the poor* has been produced by *European ideology*, as a consequence of capitalism.

In the article entitled "Nationalism returns in Europe" Nicholas Gvosdev has written that "parties of radical right are becoming *mainstream* political actors" and that sentiments of one's *dignity* and of *our state* are present throughout European states.[103] On the one hand, this concerns the strengthening of political parties, while, on the other hand, it refers to growth of a large number of separatist movements (Catalonia, Scotland, Belgium...). Finally, it also points to loud appeals in different member-states that their relation with the Union should be re-considered. So the relevance of the social issue and of the national issue, and their, in time of crisis, expanded interrelation, appears as a new demand for both *egalitarianism* and *national dignity*, as a new appeal for *equality, justice* and *dignity*, while the multiplication of repressive and control mechanisms, which tend towards new domination, reciprocally risks the turning of fear into anger as a political emotion *par excellence.*

The birth of contemporary critical European discourses of left and right, therefore – both in the sense of critical articulations and as political movements – can be analyzed through a series of discrepancies between the proclaimed and official politics of the European Union and life in European cities. Finally, the discrepancy between *EU dogma* and *European realities* reminds us of the difference between the concept of Europe and the concept of the European Union. The concept of Europe – and therefore Europe itself – exists as a *plurality, as Europes,* in a similar sense in which *democracy* is always related to *the many* and never to *the one.* "The one" always relates to totalitarianism, and precisely in this sense *the story of the EU* has presented itself *as a story without alternative.* Democratic and

---

103 N. Gvosdev, "Nationalism Returns In Europe", *The National Interest On-line,* November 5, 2012. Greek "Golden Dawn", Hungarian "Movement for better Hungary" or "Jobic", Bulgarian "Ataka", French "National Front", UK "Independence Party", "Freedom Party of Austria" or, from the side of the left, Greek Syriza, are only some of most obvious examples of this movement which can also be followed on the level of growing aspirations and attempts to *leave the EU,* i.e. in the fact that practically half of British population has this aim, while only half of Germans are willing to remain in the Union. To this list of different phenomena we can also add peculiar movements such as "Five star movement" (Beppe Grillo). In all of these cases one must distinguish between the ultra-nationalist violent movements and those which are not, since the affirmation of the role of the state cannot automatically be identified with Fascism in the same way in which appeals for social justice and equality cannot be called totalitarianism.

free Europe, on the contrary, truly recognizes otherness and therefore *per definitionem* consists of "the many". The response to neoliberal EU interventionism is called Europe, precisely in the sense in which the response to totalitarianism of the one is democracy of multiplicities.

In other words, had the architects and decision-makers in the Union preserved elements of the European traditions of freedom, justice and equality, or democratic feeling of dignity, all the stakes are that today we would not be encountering this drastic gap with still unforeseeable consequences. In this sense, Nicholas Sambanis has articulated that the European crisis is beginning to resemble, more and more, an *identity-ethnic crisis,* while Ash detected that European reality can be described as a "dysfunctional triangle" of European politics, national politics and world market – concluding that the only constant element is this European political *Rashomon.*[104]

If the contemporary EU manifests itself as an outdated conception which already, in various aspects and to the greatest extent, has been superseded, the radical left and right European politics aim at achieving, more and more, the image of alternative and progressive projects. So the question which is now placed before us is: if the rise of *radical politics* testifies to the movement of contemporary Union toward *structural differentiation,* i.e. toward legal and political fragmentation,[105] where, in this situation, is the place of *the political* and *democracy*? Or, is it rather the case that in the opposition between virtuality of the EU, and its slogan, i.e. *"the union of equal states and people", potentials for new political subjectivity* appear, moreover, the potentials which promise an eventual *end of*

---

104   N. Sambanis, "Has Europe Failed?", *The New York Times,* August 26, 2012. In this light, Sambanis reflects on the gap between "south" and "north" of Europe, i.e. the gap between the states such as Germany or France on the one hand, and Greece and Spain on the other. This gap appears as such that it is easier to imagine a potential conflict than close alliance which is suppose to establish "the unity of the people". Timothy Garton Ash claims that Germany has been particularly interested in EU integration and that "partly, it went well for Germany because it has not for southern states." In a special section entitled "European Germany, German Europe", Ash states that, in contrast to Thomas Mann's words, today we see more and more traces of "Germanized Europe" – and in it, for now, "European Germany". See T.G. Ash, "The Crisis of Europe", *Foreign Affairs,* September/October, 2012.

105   See P. Taylor, *The End Of European Integration,* Routledge, London, 2008.

biopolitics resulting, perhaps, from contemporary *master-slave dialectic*? Is it possible to say that *potentialities for democracy* and *political subjectivity* – and their interrelatedness, as *autonomy* and *self-determination* – perhaps reappear in such a situation? Moreover, if these forms of structural transformations are articulated on the level of opposition between *trans-national Euro-Atlantic imperial structures*, i.e. *EU crypto-elites*, on the one hand, and *people*, on the other, is it necessary to imagine *riots of greater extent, a series of "European Springs"*, so that we could speak of a *historical uprising and systemic change*? And finally, what is the significance of the return of the idea of *national sovereignty* in this light?

In response to these dilemmas philosophy begins from political and moral irony contained in the fact that cities such as Athens, Madrid or Lisbon, as *polis*-es in which it has been believed that there is, at least, a formal equality among EU member-states – if not a serious conviction about essential equality of different nations – have become live examples of greatest inequalities and injustice. However, the flip-side of this is that there is indisputably historical and political basis in the fact that contemporary *indignados* are becoming *indignados* in more and more European states[106] i.e. that this is becoming a *name, a political name* for those rising against *the Euro as a form of governing*. Furthermore, rising against politicians and bankers *indignados* are actually rising against EU crypto-elites – against *Eurotocracy* as the bond between the monetary governing and technocracy – and in this light they appear as heralds of new political subjectivity, the return of the political and true democracy.

So this phenomenon manifests the potentials, i.e. affirms the thesis that these contemporary processes will first be formed on the level of *transnational strikes*, as key motivators of all the more massive *European protests*. This would appear as a live expression of the fact that here we are literally speaking about a new *class struggle*[107] – in which it is precisely

---

106 The example for this is also the strike in southern Europe, i.e. "European day of solidarity and protest" in Portugal, Spain, Italy, Greece and Cyprus, as well as the continuity of protests in different countries directed against EU politics.

107 S. Wagenknecht, "The Strike In Southern Europe", *Global Research*, November 12, 2012.

the slave who opens the scene of *new history*. On the other hand, one should certainly not underestimate the relevance of the national issue that, in this situation, no less concerns the questions of freedom and justice, practically, therefore, falling into one with the social phenomena. Because, from the aspect of political sovereignty or, more accurately, *popular sovereignty*, consciousness about the relevance of particular European states is growing.

Such a space, of resistance and subjectification, today has been open, but the realization of such a possibility on philosophical and political grounds is proportional to the rise and participation of citizens and to the development of political self-consciousness i.e. to subjectification and re-politization. How these processes correspond to Mouffe's writings, we can see by remembering that in *The Democratic Paradox* she has articulated that it has been a strategic mistake that the idea of popular sovereignty has been erased from theoretical and political discourse. Moreover, precisely from the perspective of interrelatedness between radical democracy, popular sovereignty and democratic institutions, Mouffe has reflected the contradictory nature of the so-called *liberal democracy*: in the attempt to reconcile completely different traditions of individual freedom and human rights (*liberal tradition*), and equality and popular sovereignty (*democratic tradition*), liberal democracy has been caught in its own structural tensions. *Liberalism is not democracy* and, therefore, the reduction of *democracy* to *liberal democracy* – that for its consequence had equalizing *democracy* with *procedure* – has been the cause of ambivalences and internal contradictions.

Furthermore, when democracy is identified with procedure and with formalities such as voting on elections or competition of seemingly different political options which are a part of the same system – and this shows what happened with simulation of politics of left and right, with fake politics – political power is handed out to self-proclaimed elites, and the entire public sphere, the sphere of debates, arguments, common action of citizens is erased. The logical consequence of this has been the growth of abstentions in elections, apathy and passivity in regard to so-called normal governmental politics, because one of the best and oldest European traditions, the Aristotelian action of human being as political being – that is being that has reason (*zoon politikon* and zoon logon echon) has been brought into question. However, these processes have shown that *rethinking of democracy* and *political subjectivity* must take an entirely new direction – and most likely one which brings it closer to its original concept. For when theories are refuted empirically, then it is time for the birth of a new theory.

In this light, therefore, the opposition to what Badiou called *"capital-parlamentarism"* – which refers, above all, to the usurpation of political sovereignty and economic sovereignty, i.e. to the entire field of popular sovereignty occupied by crypto-elites – appears as the strengthening of *subjectivation* in different aspects of personal and collective existence; in the sphere of *oikos* and in the sphere of *polis,* as well as in divergent aspects which point to *authentic action.* This is plausible for Greece, and it was articulated in the appeal of European intellectuals "Save Greece from its saviors!"[108] – which implies the *return of the state to its own potentials and the return of political and economic participation of citizens in its formation.* In other words, it means the reaffirmation of *autonomy* of all *inexistent,* which, in time, as the so-called *"transitional losers",* have actually become *the majority.* This is why the struggle for truly *social Europe* of justice, freedom and equality, has as its *sine qua non* the struggle against the position of being on the margins of the Euro-Atlantic Empire In such a sense it also signifies the struggle for the return of *national and personal dignity.* More precisely, to the extent to which it grows as a struggle for *equality,* on various levels marked by contemporary French cry *"Casse-toi, riche con!"*, this struggle, at the same time, appears as a *systemic riot* and, as such, is theoretically and politically formed through different, and entirely new, *critical European discourses and practices.*

---

108 See A. Badiou, J.C. Bailly, E. Ballibar, C. Denis, J.L. Nancy, J. Ranciere, A. Ronnel, "Save The Greeks From Their Saviors", www.europeagainstausterity. org.

Time Of Small Revolutions, Biopolitics And True Democracy

Different contemporary phenomena we have turned to here – the new world images and manifestations we have attempted to articulate and conceptually define – the Arab Spring, the Occupy Movement and the birth of radical left and right European politics – suggest, in a multiplicity of aspects, some major questions. Is the twenty-first century a proper *biopolitical time* or, conversely, are the silhouettes and shades of the *end of neoliberalism* appearing in our epoch? Or, more explicitly, if we see the first traces of the manifestation of *political subjectivities* on the horizon, is this enough to suppose that the ruptures in the system have grown to the extent that one can speak of an irreversible process?

Prevailing theoretical and practical conditions suggest that, *en generale*, we still live in the world of biopolitics. Because, in *contemporary biopolitics,* on the level of *technology* and *sexuality everything is possible and everything is allowed* in all spheres – so long as it does not infiltrate the heart of *politics* and *economy* and thus threatens the *system* itself. This is to say that *political virtuality,* as well as the simulation of economic sustainability of structurally unsustainable paradigms, with expansive roles of *crypto-elites*. still enables the *hyper-production* of the *status quo* in the Western and some parts of the non-Western world. Through multiplicity of techniques of power, domination and intervention, in Arab countries the influence of imperial and colonial forces is still present and visible; the OWS movement, at this moment, is not going through its best days, while, on the third side, in spite of numerous new processes occurring on European grounds, the dominance of force is still at stake, even when most of *the people* are opposed to it.

However, in spite of the fact that this theoretical-practical crime of *virtuality of the oikos and polis* presents itself as practically a *perfect crime* – which has caused systematic production of *de-politization* and *de-subjecification* and therefore conditioned neoliberal control of the entire populations by *trans-European crypto-elites* – all the stakes are that this time can be named *the time of intervals. The time of intervals,* as a particular time gap, represents *the path between two epochs*, the *topos* in which the material on the basis of which a transformation will occur is gathered, collected and overturned. It is a time in which multiplicity of things *appear and disappear,* coming out from the blue or vanishing without a trace; time of stories but equally of confusion and unclear hope – *time of small revolutions.*

As such, it is different from the time of *grand revolutions* which signify *the event, the gap and a new epoch.* Because the *time of small revolutions* – initiated by the Arab Spring, the Occupy Movement and the European politics of left and right – for the most part is a *time of potentiality, multiplicity of created occasions and small movements.* If in this *interval,* in the interrelation between the political-philosophical and the historical forms, and contingent circumstances of the pulsing of social bodies, *new ethics* and *new politics* come into being, we can speak about an irreplaceable ideological turn which establishes a *new system. Ergo, the time of small revolutions can* – but does not have to – represent the beginning of the *time of grand revolutions,* since, like the difference between *immediate riots* and *historical uprisings,* it can appear as the *voice of a new era*; if it turns out that it, at least as an announcement, refers to tectonic transformations, political, social and economic earthquakes after which nothing remains the same. So the *time of intervals* is not – in the sense in which Hardt and Negri spoke of the *multitude* – some collateral superior excess which the system produces against itself, but, on contrary, it is a signal and a symptom of crisis and eventual end of processes of *contemporary biopolitics.*

Indisputably, precisely as chaotic, limited in space and time, *prepolitical immediate uprisings* can end in mere and naked *destruction and/ or self-destruction,* or simply *end as if they had never happened.* Equally *the time of intervals* can finalize in the establishment of infinite regress, *the other of the same* which seems to last *ad infinitum,* i.e. with a system which will maintain all relevant tools for *biopolitical hegemony.* As such, it will not signify a *qualitative leap* into *the political* but will rather present itself as the *continuity* of its *mimesis.* In this light, the condition for the affirmation of the appearance of a new epoch refers, first, to a broader existential transformation on all levels of *subjectification* and to the multiplicity of *intersubjective relations* both in the public and the private sphere, i.e. it refers to the *emancipatory turn* which places forward simultaneous transformations in the field of politics, science, art, culture and media – manifesting *the normative and structural change.*

In this sense, it is worthwhile to remember that *democratic tradition* – or, more accurately, the tradition from which *the political* comes forth precisely through the questions of *popular sovereignty* and *equality* – has found its particular expression in the concept of true democracy (*Wahre Demokratie*), as a movement which refers to *living democracy*, i.e. to a *true and real dimension* of *the political as democratic action.* This means that what one encounters here is the interweaving of an epistemological moment and the moment of *praxis*, and that, moreover, the symbiosis between the theoretical and the political, appears in the question of *the people* and its *political subjectivity.* Marx's critique of Hegel's *Philosophy of Right* is an exemplary case of the attempt to bring into daylight the sense of *true democracy*, as an irreplaceable type of *commitment* and *loyalty* to the conception of *the government of the people*, and to the process of *subjectivation of the people*, namely, how people become a subject through self-governing.

In 1843 Marx referred, simultaneously, to two relevant aspects, let us say to a *subjective* and *objective* moment of this process. The first refers to *personal and individual faithfulness to the democratic moment* – which emerges as an echo in Badiou's concept of *"faithfulness to the event"* – while the second is related to the process of *grounding and preserving the community.* This is why Marx's claim that it is self-evident that all state forms have democracy as their truth and are therefore untrue to the extent to which they are not democracies – appears as a fundamental insight which perceives the issue of *democratic self-determination*, that is of *democratic self-founding* (*Selbstbestimmung des Volks*), as the exemplary feature, as the *topos* of *the political.*[109]

In difference, therefore, to all procedural *ergo* formal theories of democracy, that belong to liberalism – philosophical and political conception of *true democracy* refers to a structural interweaving between *the democratic* and *the political* and, in the ultimate sense, argues that *the question of democracy* is *the question of the political* and *vice versa.* Because what is at

---

109   K. Marx & F. Engels, *Werke*, Karl Dietz Verlag, Berlin, Band 4, pp. 464–5.

stake is precisely *the autonomy of politics,* the specificity and particularity of the political in relation to practically all other forms of human activity: because only *politics* is the event of *self-constitution of the people* and inasmuch as in reality this is the case – *it is true as politics.*

Furthermore, a *state,* as well as all *state forms,* for Marx are *true* – therefore, conformed to their concept – if they appear *as democracy,* and the level of democracy in a state is practically proportional to its structure, strength and sustainability. So, on the one hand, the question of democracy turns out to be a question of political legitimacy, and, on the other – coming from Hegel's difference between *Schein* and *Erscheinung* – it appears as the difference between *mere illusion* and *relative truth.* Marx develops the idea of the gap between *form* and *reality* and does so in such a way that *the level of realization of democracy* immediately reflects *politics* and *anti-politics.*

Moreover, true democracy is equally the departure from *virtual worlds,* from a media- constructed world image, *as a return to society of knowledge,* of education, of *paideia,* and an enlarged participation of citizens in political and public life. Therefore, true democracy appears as a specific imperative, that contributes to the development of what Ali recognized as current great weaknesses – *critical thinking* and *political imagination.* True democracy – the struggle for true democracy – is, above and beyond all, the matter of *the people* and of *the public sphere,* of *politics as public activity* opposed to crypto-political bureaucracies and technicians of power.

However, true democracy, first and foremost, refers to overcoming the gap between the *crypto-political class* and *the people.* Its departure point lies in the perception of citizens that their so-called political class is *not theirs* – for the interest of quasi-elites excludes the people. Secondly, true democracy relates to the recognition of illusion about the beauty of the market, i.e. it relates to understanding that the liberal-democratic or, rather, neoliberal model, is the cause of the economic and social gap (for the rule of such a system is the rich getting richer and the poor getting poorer). Third, as a return to *knowledge,* the new democracy goes against *knowledge as information,* and against education as technology, management and hyper-production of *experts* in non-existing fields. This is because educated citizens are even more capable to politically decide and govern themselves, that is, to *act as political beings*, while social consciousness, as well as science, art and culture are realized on a higher level.

On the other hand, this does not mean that anyone is always better in political matters than anyone else, for, as Ranciere has articulated, true democracy means that in principle, no one is more capable of governing than anyone else.[110] The only criteria for reaching a political decision is the strength of arguments placed forward, and the only space for the political is a free debate in public. Simultaneously, this is a call to remember that politics is always – a public matter and a matter of the public. True democracy refers, also, to the new movement of political imagination, to the task of inventing new forms of political and social action – and therefore to a new potentiality of *politics as power* (or, equally, of *the concept of power*). The break with the idea that democracy is simply electoral democracy may be the beginning of a new world, the horizon in which the time of past and future emerges – replacing the neoliberal demand for the eternal present.[111] Because, the element of *the will of the people* is what *disturbs the system* and the opponent to *universal monarchy* is *democracy*.

We could here recall Plato's myth about the cave, the ironical twist that concerns true knowledge: trapped in the cave, *as spectators of shadows*, the people first see only *shadows of reality*, while in stepping into *sunlight* they begin to see everything else. Also, this turn to *vision of reality* philosophically can be traced as a return from Hobbes to Rousseau, from the condition of war of all against all, individualistic egoism and solipsism, to rethinking of the community. In difference to Hobbes's concept in which sovereignty reflects the power of the one who governs, Rousseau's idea is

---

110 See J. Ranciere, *Dissensus*, Bloomsbury Academic, New York, 2010.
111 Neoliberalism revealed itself as a specific *control of time* which, first and foremost, excludes the past and the future. Post-subjectivity in the materialistic paradigm has been constructed through the dominance of an "*I*" and "*now*". In the project of the eternal present both normative and critical dimension of human existence i.e. of subjectivity in horizon of time have been ruled out. Consequently, the conclusion that followed from the proclaimed end of history has been that nothing new and different can happen and this is how the time of future was erased, together with the whole sphere of practical activity and the sphere of the event. The biopolitical demand for the eternal present rests in the idea that future has already happened and that the past never existed.

that the subject of sovereignty is the people. This would, therefore, be a transition from a decesionistic model of sovereignty (where the power has *the one who decides*), to popular sovereignty (as soon as there is a *master the body politic* is destroyed), i.e. to the creation of the general will and its realization, which refers to the event of the political.

Rousseau's thinking of the political is precisely what neoliberalism cannot tolerate, because the core of his idea has been that the ultimate carrier of sovereignty is *the people,* therefore, the perspective which opens *the horizon of the political.* And, coming back to what we already said about *invisibility and visibility of power*: *invisible character of power* is proportional to its *crypto-political success*,[112] while the moment when it is forced to *step into light* represents the moment of its *weakening,* that is, the moment in which *dialectics of power*, therefore, *resistance,* begins to *constitute itself.* The moment when biopolitics – or rather, different phenomena of contemporary biopolitics – had to enter *the public realm* is the moment which reveals *the absence of the subject, the actual subjects of power,* and, further, it is the moment in which *perspectives for alternatives rise.* In other words, because *the invisible power* has been directed to fragmentation of political and economic sovereignty, and to undertakings which would make democracy senseless, *the will of the people* appears as the key factor of disturbance which can potentially overturn the system. This opens Badiou's horizon of *the event,* the sphere of *the incalculable,* from which the creation of new beginnings – such beginnings which contain the consciousness of *praxis* – rises.

Is it, therefore, possible to imagine that the quest for true democracy could primarily be realized in the forms of strikes, as motivators of protests in the world, something that could be a live expression of *contemporary class struggle*, of a *new proletariat?* Will millions of those who are hungry be the prime mover in a drama in which Hegelian master-slave dialectic will enter the scene? Because the final outcome of this dialectic presents the fate of oppression by its concept: in the moment when the slave gains

---

112 The process of *depolitization of the world* has been exemplified by *power* which lies in the sphere of *invisibility* in spite its brutality in the economic, the political and the social domain. As in Baudlliard's description of *the perfect crime* the subject of *absolute capitalism* remained concealed.

consciousness about his position, and becomes aware of the fact that he does not have to be slave, the master-slave relation overturns completely. In that sense, the first condition of realizing freedom in the outer world is recognizing its possibility, and when this condition is fulfilled the perspective of resistance rises. Is the fact, therefore, that the *struggle for recognition* is becoming not only struggle for *dignified life* but for *bare life*, is that fact the first element of self-consciousness and are we currently reaching the stage in which *struggle for life* and *struggle for freedom* appear as inseparable? Is the point in which the question *to be or not to be* becomes an existential issue the final step of creation of resistance? In this issue we see relevance of Marx's analyses of the relation between the political and *the economic* in capitalism and in its alternatives.

Certainly, the look and the outcome of these contemporary processes will depend on series of circumstances and precisely because this belongs to the category of the event, this cannot be calculated. What is relevant to keep in mind is that the common ground for action lies in the call for the end of neoliberalism, the end of biopolitics and interventionism i.e. in the call for the autonomy of politics and for *the role of the people as the political subject* – as the call for *true democracy*. Moreover, in the example of rethinking, let us say, new Europe after the EU, true democracy can, for instance, go in the direction of creating forms of *local autonomy*, of self-government in particular "spaces of life". This can be a basis for a new political system, where people *represent us,* because it is we who *represent ourselves.*

As the alternative to the myth of EU, we could speak perhaps of *communitarian confederation of European states* – as the alternative that grows in the heart of Europe. Furthermore, this could be a project of *cosmopolitan communitarianism, the project of Europe of free and equal people, where real differences are preserved* – a potential cosmopolitan community of people and cultures, of multiplicities. We could also think of at a *new Europe of cities* or – in expansion of this concept – a *new world of cities* instead of *megalopolises,* which all look alike, and look as all other megalopolises in the world. In this new Europe, *cities* such as Paris, Rome or Athens would again shine in unique and irreplaceable forms of their cultures, in differences of infinite multiplicities as the true content of European culture.

The bottom line is that communities could be based on *new social contracts* and that emancipatory ideas could emerge as the *telos* of action. Such action can be practiced only in a communitarian dimension, and of *local character,* as practice of concrete communities which articulate the social bond of solidarity and anti-biopolitical undertakings. This way the gap between *the political class* and *the people* can be overcome and citizens would regain the right to decide on their own destiny. In other words, *popular sovereignty* is the best guarantee for the existence of *communitas.* This is the bond between *true democracy,* and ideas of *justice, freedom and equality.* These are the reasons that enable us to say that time of transition from one epoch to a next one, the time between two epochs, is rising as the perspective for new beginnings – *the time of the political, the return of the political and its creation.* This is, no less relevantly, *time of knowledge as well,* the departure from the eternal present of neoliberalism, that aimed to produce a market society in which there would be no place for non-profitable activities.[113]

The fact that philosophy preserves the utopian moment – in the sense in which it refuses to be reduced to the level of *Realpolitik* as exclusive reality – enables its revolutionary character in thinking, i.e. its capability to produced new worlds and completely different logics. In time of biopolitics the first task of thinking is the articulation of strategies of resistance and, secondly, the creation of various forms of social gatherings – as the movement towards *the time of the political.* Opening the horizons of past and future, philosophy discloses the infinite field of possibilities, reminding us that what we can imagine is still not all and, no less relevantly, that the dialectics between the universal and the particular is the point of emergence of truths. Also, here we can recall Foucault's statement that philosophy in

---

113 This anti-philosophical, anti-cultural and anti-civilizational impulse of biopolitics comes from the fact that the beginning of political and social resistance lies in thinking, i.e. in free critical consciousness interested in truth. Consequently, as the level of theoretical and practical colonization is higher, so is the role of philosophy as subversive activity. In this light, Hegel's insight that every single individual is capable of philosophy corresponds to Ranciere's conclusion that all people, in principle, are capable of governing. Philosophy is democratic in character i.e. as true democracy it is a public matter and a matter of the public.

our time is dispersed to different activities: the linguist, the ethnologist, the historian, the revolutionary, the politician are roles which can present themselves as philosophical *par excellence*, and contemporary philosophical engagement can take on many particular forms. Because in order to exit the condition of *false present* – which is false by the very fact that it has been *totalized* – the return to *the horizon of time*, i.e. the return to *the question of the truth* includes in itself an extensive field of theoretical and practical action. The return of time, i.e. the return of three time dimensions of being reopens therefore the question of truth. And then it is possible to say that *the time of the political* has come.

Subjectivity and *time* are interrelated in the most fundamental way i.e. any form of individual or collective subjectivity is realized in *time* and *vice versa*, time without a subject ceases to exist. In this sense, different "deaths", the death of the political, the death of subject and death of history appear as results of the attempt to *end time*. Moreover, this is why it is necessary to rethink transformation processes in *time* and *space*, because the first condition of an *uprising* is constitution of *the time of the political* as time in which there is reflection on time, i.e. on history and creation of the future. Furthermore, the time of the political, which always includes in itself invention and freedom of practical action – opening the sphere of the event – also includes *a particular space* in which it takes place. This is because systematic transformations occur in specific *topos* or, rather, in multiplicity of them, and in this way they co-belong to the event of the political. This type of interweaving between subjectivity, time, space and the political presents the basis of birth of new worlds. And this is the live image of a *time of historical riots*, in the sense in which uprising presupposes the emergence of a new clear transformative idea, and people who are ready for its practical realization.

Badiou has recognized that, in order for a *riot* to be *historical* it must expand to practically all parts of population and especially to those which, regarding their status, social place, gender or age are far from its *constitutive centre*. This argument appears as a test not only for the events in the Arab world but also for the events in the heart of the Empire, as well as on the European continent. Because it is precisely the *"qualitative extension"*, i.e. a collection of divergent and mutually opposed parts of population – the young, workers, intellectuals, families, women, even police officers and soldiers – that makes the *multiplicity of voices* fall into the unity of *the people*. Moreover, it is only then that we can legitimately speak about *people* becoming *a subject* and about the birth of *the political* which on the horizon appears as *the democratic*; as the authentic *government of the people*, as autonomy, self-positing, *(self)critique* and *(self)consciousness* which accepts the *responsibility* for *historical decisions*.

This is the reason why, at the *time of small revolutions*, besides relevance of *space* and *duration* – *topos* becomes relevant because it *always* refers to what is already *historical* whereas location appears as significant precisely because of its *"extension"* which becomes a feature of decisive importance when the *beginning of the foundation of a new subjectivity* appears. And since the *time of intervals* can sometimes last for decades or even longer, perhaps decisive for the movement into a new epoch is the formation of *bonding*. Such bonding, of the specific *material,* is between what is yet *unknown*, between the *unnamed*, but nonetheless pervades all spheres of being and expresses the peculiarity of a riot.

The *time of intervals*, therefore, can potentially be fulfilled with series of *protests, local small or bigger riots and strikes, numerous small-scale social turbulences, different forms of civil disobedience,* simultaneously as it can, on the other hand, be remembered by the *political, legal* and *economic interventionism* on domestic and foreign grounds. This means that, in the fluctuant time gap, in which one system disappears while a new one is not yet manifested, it can happen that *institutions, laws, rules, social and normative expressions, even entire populations, people and states* nearly over night, and sometimes even several times, *transform their image to the extent of becoming unrecognizable,* because *the fluidity* which is the basic content of *the time-space gap* often carries with itself *the contradictions and the clashes* – but also *a flow of events which cannot be calculated, which represents another announcement of the political, i.e, the announcement of exit from the system of a perfect domination and control.*

In this sense, the transfer which refers to the first ruptures in the liberal system of governing can be noticed in *all three different contemporary phenomena* which we have discussed. For the *Arab Spring* it is the question of the *time after,* i.e. the time in which, eventually, will prevail those convictions which do not serve the Empire but rather oppose it For the Occupy Movement and the parole *"We are the 99%!"* it means the direction and the intention for the *structure of whole neoliberal system* and unclear, but still present appeal against *political representation per se*. For the contemporary European politics of left and right the most important feature is *the radicality* by which they embody the *sense of the concept of the political*. Or, to the extent to which in the *non-Western world* – which has been practically completely colonized – *new ideologies*, based on *equality, justice and freedom,* are being articulated, the potentialities of *time of small revolutions* appear.

At the same time, in a minimal but relevant sense, traces of resistance in the heart of biopolitical system, i.e. in the West, which structurally manifest serious ruptures, refer in a similar way to the possibility of *leap* to a new epoch. And finally, the birth of contemporary critical European discourses shows that for Europe which would, eventually, return to itself and its best traditions, *there is still some time left*: knowledge that the EU is not only an *anti-political* but equally an *anti-theoretical* project, is the first step in this direction. Certainly, only when, in all the three examples, we begin to notice *the return of the state to its own potentials, i.e. the return of political participation and participation of citizens in political life (new politization* in the sense of the rebirth of *polis)*, the *creation of different and diverse economic models* – and finally, but no less significantly, the reaffirmation of *the autonomy* of all the *invisible* and the *inexistent* and their *subjectivation* – we will be able to say that the *time of intervals* in the twenty-first century has turned out to be time of *historical uprisings*.

Because in such a case, the political-philosophical sense of *the event* – such as the thought that all *"transitional losers"*, as liberal capitalism has named them, and which have in time become *the majority*, appear as *carriers of political subjectivities* – will turn out as achievable and *desirable*. So the turbulent time of the contemporary world is the force which can *overturn* biopolitics, and with all presupposed risks which struggle for justice and freedom carries with itself, the *time of small revolutions* can be its *authentic and irreplaceable beginning*. And if one thinks that only what exists is changeable – then what exists is not all.

BEYOND HUMANISM: TRANS- AND POSTHUMANISM
JENSEITS DES HUMANISMUS: TRANS- UND POSTHUMANISMUS

Edited by / Herausgegeben von Stefan Lorenz Sorgner

www.peterlang.com

BEYOND HUMANISM: TRANS- AND POSTHUMANISM
JENSEITS DES HUMANISMUS: TRANS- UND POSTHUMANISMUS

Edited by / Herausgegeben von Stefan Lorenz Sorgner

www.peterlang.com

www.ingramcontent.com/pod-product-compliance
Lightning Source LLC
Chambersburg PA
CBHW031542260326
41914CB00002B/224